T0275662

MODELING OF CHEMICAL WEAR

MODELING OF CHEMICAL WEAR

Relevance to Practice

A. SETHURAMIAH

Former Professor
Indian Institute of Technology Delhi
New Delhi, India

RAJESH KUMAR

Associate Professor
Department of Mechanical Engineering
Indian Institute of Technology (BHU)
Varanasi, India

ELSEVIER

AMSTERDAM • BOSTON • HEIDELBERG • LONDON • NEW YORK
OXFORD • PARIS • SAN DIEGO • SAN FRANCISCO • SINGAPORE • SYDNEY • TOKYO

Elsevier
Radarweg 29, PO Box 211, 1000 AE Amsterdam, Netherlands
The Boulevard, Langford Lane, Kidlington, Oxford OX5 1GB, UK
225 Wyman Street, Waltham, MA 02451, USA

Copyright © 2016 Elsevier Inc. All rights reserved.

No part of this publication may be reproduced or transmitted in any form or by any means, electronic or
mechanical, including photocopying, recording, or any information storage and retrieval system, without
permission in writing from the publisher. Details on how to seek permission, further information about the
Publisher's permissions policies and our arrangements with organizations such as the Copyright Clearance
Center and the Copyright Licensing Agency, can be found at our website: www.elsevier.com/permissions.

This book and the individual contributions contained in it are protected under copyright by the Publisher
(other than as may be noted herein).

Notices
Knowledge and best practice in this field are constantly changing. As new research and experience broaden
our understanding, changes in research methods, professional practices, or medical treatment may become
necessary.

Practitioners and researchers must always rely on their own experience and knowledge in evaluating and
using any information, methods, compounds, or experiments described herein. In using such information
or methods they should be mindful of their own safety and the safety of others, including parties for whom
they have a professional responsibility.

To the fullest extent of the law, neither the Publisher nor the authors, contributors, or editors, assume any
liability for any injury and/or damage to persons or property as a matter of products liability, negligence or
otherwise, or from any use or operation of any methods, products, instructions, or ideas contained in the
material herein.

ISBN: 978-0-12-804533-6

British Library Cataloguing-in-Publication Data
A catalogue record for this book is available from the British Library.

Library of Congress Cataloging-in-Publication Data
A catalog record for this book is available from the Library of Congress.

For Information on all Elsevier Publishing publications
visit our website at http://store.elsevier.com/

Working together
to grow libraries in
developing countries

www.elsevier.com • www.bookaid.org

CONTENTS

PREFACE

This contribution is the outcome of few years of collaborative effort between me and my younger colleague Dr Rajesh Kumar. His contribution to chapters 6—8 is significant. Part of the present book is based on my previous book LWST (Lubricated Wear—Science and Technology, Elsevier, 2003). However the present volume is so written that it is self-sufficient. The scope and purpose are clearly explained in the back cover.

I wish to thank my wife Jayashree for all her support during this endeavor.

I fondly remember all my former students who contributed directly and indirectly to this book.

My special thanks are due to Dr G. N. Rao, former professor of chemistry, IIT Delhi for clarifying certain aspects of chemical hardness.

I stay abreast with current developments thanks to my younger colleagues Professors S. Kailas and M. Bobji of IISc Bangalore. We have periodic discussions on tribology with students that are exciting. These interactions have an influence on the tenor of this book.

Throughout Elsevier staff have been very helpful and we wish to thank all of them for achieving the well designed final book.

A. Sethuramiah

At the outset special thank goes to my mentor Professor A. Sethuramiah for motivating to participate in this challenging task which covers briefly the entire tribology, overview of vast areas of DOE and optimization along with examples and computer programs relevant to industry.

I wish to thank Professor R. Sangal, Director, IIT(BHU) Varanasi for granting me sabbatical.

I gratefully acknowledge Professor R. B. Rastogi, Department of Chemistry, IIT(BHU) for valuable discussions in the area of tribochemistry.

I would like to thank my former PG students H. Saini and R. Rabha whose works find a place in Chapter 8.

My thanks are also due to doctoral students V. Jaiswal, and Ambuj Sharma; PG student Mayank; UG students Amit, Rajat, Pushpam, Ayush, and Madhup for helping in graphics, nomenclature and computer programming.

Last but not least I would like to thank my wife Dr Vinita, sons Vivek and Anand for their support, care and patience during completion of the work.

We hope the readers will find this book useful and interesting.

Rajesh Kumar

CHAPTER 1

Tribology in Perspective

1.1 INTRODUCTION

Tribology is defined as the science and technology of surfaces in relative motion. It encompasses the well-known areas of friction, wear, and lubrication. The three components are interconnected and need to be studied together. To emphasize this fact the word tribology, based on the Greek word *tribos* (which means rubbing), has been coined.

Conservation of energy and materials is the more obvious facet of tribology. Knowledge of tribology is necessary both in selecting the materials and coatings as well as their evaluation. In some cases the tribological problem is so vital that unless it is solved new technology cannot be implemented. An interesting example is the adiabatic diesel engine. This can be achieved only by developing high temperature wear-resistant ceramic materials, and is not yet successful. Another example at a micro level is micro electro mechanical systems (MEMS) technology. In MEMS any sliding between silicon surfaces results in drastic wear. Hence MEMS technology is now restricted to components that do not slide. In such cases defining tribology as an enabling technology is appropriate.

The first section in this chapter considers surface roughness and contact. The known laws of sliding friction and their explanation based on plastic contact are considered next. This is followed by modeling elastic contact and its relevance to friction. Here a new model based on our work is also presented to calculate the contact area of non-Gaussian surfaces. The limitations of the model are then discussed, followed by a model that considers the special case of growth in the contact area. A brief consideration is then given to the modern approaches to modeling of friction at the micro and nano levels. The final section deals with lubrication. The concepts involved in hydrodynamic and elastohydrodynamic lubrication (EHL) are explained, followed by mixed and boundary lubrication. In view of its relevance to the present book, boundary lubrication will be considered in detail in a later chapter.

Modeling of Chemical Wear.
DOI: http://dx.doi.org/10.1016/B978-0-12-804533-6.00001-9
© 2016 Elsevier Inc.
All rights reserved.

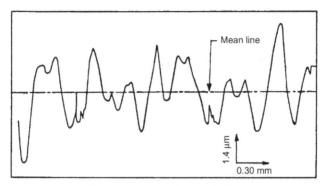

Figure 1.1 Roughness profile of cast iron surface with $R_a = 0.88$ μm.

1.2 SURFACE ROUGHNESS CONTACT AND FRICTION

1.2.1 Roughness

Every engineering surface is rough and has fine undulations. Shorter wavelength height variations around the mean line that typify the manufacturing process is referred to as roughness. Surface may also have longer wavelength variations, referred to as waviness. In such a case the longer wavelength influence is filtered off to evaluate roughness. The roughness values of surfaces can vary from 0.05 to 5.0 μm. A typical roughness profile of a cast iron surface is given in Figure 1.1 as traced by a stylus instrument. The high points are called *asperities*. They have a gentle slope but appear sharp in the trace. This is because the vertical magnification is much higher than the horizontal magnification. The trace can be analyzed for different parameters based on digitized data. The commonly used parameters are R_a the center line average, and R_q the rms value. These are based on the mean line that divides the surface into equal areas. The other parameters commonly obtained by the profilometer include number of asperities per unit length, asperity slope, and curvature. An additional parameter that is used for many surfaces is R_z, which is the mean distance between the five highest peaks and five lowest valleys. This parameter is used to characterize the sharp peaks on the surface. The values obtained are two dimensional along the chosen line. Several other parameters are derived and used for various purposes. Only normally used parameters are considered here.

1.2.2 Contact of Rough Surfaces

The contact between a rough surface and a rigid flat is considered first. This will then be adapted to the situation of contact between two rough surfaces, as considered later.

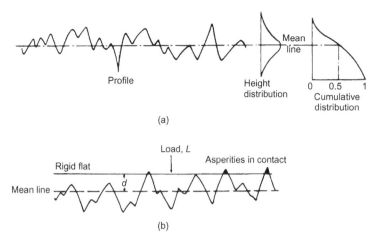

Figure 1.2 (a) Roughness trace and Gaussian and cumulative height distribution. (b) Loaded contact between rigid flat and rough surface with separation d.

The contact situation is given in Figure 1.2. The nature of contact will depend on the nature of asperity distribution. If this distribution is Gaussian it may be characterized as shown in Figure 1.2a. When this surface is brought into contact with a rigid flat under load the contacts will occur at few peaks as shown in Figure 1.2b. Hence if d is the distance between the mean line and the rigid plane, all the asperities whose height is greater than d will come into contact. At any asperity with height z_i that is in contact the deformation δ_i will be equal to $(z_i - d)$.

From this consideration it can be seen that real contact is at a few spots and much smaller than the geometric area. The sum of asperity contact areas is called the *real area*. Two possible situations may be now considered. In one, the stresses at the spots are high enough to yield plastically. This is considered under plastic contact and forms the basis to explain laws of friction. The elastic contact situation will be considered next.

1.2.2.1 Plastic Contact and Amontons' Laws of Friction

Let us assume that the contact spots are plastically deformed and distributed over the nominal area. This is illustrated in Figure 1.3.

In this case all spots are subject to the plastic flow. The flow pressure is given by hardness H of the softer surface. The reason for using hardness instead of usual yield strength σ_y is because circular contacts attain full plasticity when the mean pressure reaches that of hardness. According to Hertz theory, the yielding starts at the subsurface beneath the center

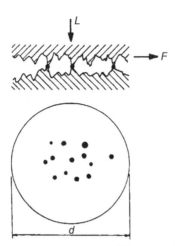

Figure 1.3 Contact spots within apparent area of diameter d.

of circular contact when mean pressure reaches $1.07\sigma_y$. At this stage the plastically deformed material is surrounded by the elastic material. The deformation is elasto-plastic. As mean pressure increases the plastic zone grows until it reaches the surface when mean pressure becomes equal to hardness. The Hertzian stress distributions are considered in Chapter 3.

Explanation of laws of friction normally attributed to Amontons was first based on the preceding contact model. The laws due to Amontons are stated below:

1. Friction force is proportional to normal load.
2. Friction force is independent of the geometric area of contact.

These laws have been explained by Bowden and Tabor [1] on the basis of the plastic contact model described earlier. In view of the assumed plastic contact at asperities the normal stress acting at any asperity is equal to the hardness. Also they assumed that the shear stress needed to break the junction is equal to the shear strength of the material. The junctions are called adhesive junctions and the theory is generally known as the adhesive theory of friction. Here the word adhesive is used with regard to micro junction formation akin to cold welding. The following expressions may now be written:

$$A_r = \frac{L}{H} \tag{1.1}$$

$$F = s_m A_r \tag{1.2}$$

$$f = \frac{F}{L} = \frac{s_m}{H} \tag{1.3}$$

where

A_r = real contact area (m²),
L = load (N),
H = hardness (N/m²),
F = friction force (N),
s_m = shear strength of the metal (N/m²).

According to Eqn (1.3) friction coefficient f is a constant since s_m and H are material properties. It follows that friction force is proportional to load. As real area is governed only by hardness and load it is independent of geometric area. Hence friction force is independent of geometric area.

This explanation is based on contact of metallic materials. When the contact pair is not identical it is considered that the friction is governed by the properties of the softer material.

1.2.2.2 Elastic Contact and Friction

The asperity contacts for many engineering surfaces may be elastic, elasto-plastic, or plastic. In such cases it is of interest to know what the real contact area is and how it affects the laws of friction.

The available theory for elastic contact is now considered. Refer to Figures 1.2 and 1.3 and replace plastic contact with elastic contact. For elastic contact, the mathematical formulation of Greenwood [2] is given below. In this formulation each asperity contact is governed by Hertzian theory.

$$n = \eta A_n \int_d^\infty p(z)\mathrm{d}z \tag{1.4}$$

$$A_r = \pi \beta \eta A_n \int_d^\infty (z-d)p(z)\mathrm{d}z \tag{1.5}$$

$$L = \frac{4}{3} E \beta^{0.5} \eta A_n \int_d^\infty (z-d)^{3/2} p(z)\mathrm{d}z \tag{1.6}$$

where n is number of contacts, A_r is real area, L is Load, z is asperity height, d is separation, A_n is geometric area, η is asperity density, and $p(z)$

is the asperity height distribution function. In Figure 1.2, the mean line is based on general height distribution while in the present case the mean line will be based on asperity height distribution.

It is convenient to use standardized variables, and describe surface heights in terms of nondimensional number $h = d/\sigma$. Here σ is the standard deviation. On this basis Eqns (1.4)–(1.6) reduce to

$$n = \eta A_n F_0(h) \tag{1.7}$$

$$A_r = \pi \beta \eta A_n \sigma F_1(h) \tag{1.8}$$

$$P = \frac{4}{3} E \beta^{0.5} \eta A_n \sigma^{1.5} F_{1.5}(h) \tag{1.9}$$

where

$$F_n(h) = \int_h^\infty (s - h)^n p(s) ds \tag{1.10}$$

$p(s) =$ standardized asperity height probability density function.

When these equations are solved assuming Gaussian distribution, real area is nearly proportional to load. In the preceding theory, β is average asperity radius and E is modulus of elasticity. When two rough surfaces are in contact it is treated in terms of equivalent surface against a rigid flat. The modulus of elasticity, rms roughness, and asperity radius of this equivalent surface are derived as follows:

$$\frac{1}{E} = \left(\frac{1 - \nu_1^2}{E_1} + \frac{1 - \nu_2^2}{E_2} \right) \tag{1.11}$$

where subscripts 1 and 2 refer to the two surfaces.
    Also,

$$\sigma = (\sigma_1^2 + \sigma_2^2)^{1/2} \tag{1.12}$$

$$\frac{1}{\beta} = \frac{1}{\beta_1} + \frac{1}{\beta_2} \tag{1.13}$$

These equations are based on the individual asperity undergoing elastic deformation on 2D representation of surfaces. Designating high spots over the surface as summits and those in the profile as peaks, the following assumptions may be made for a Gaussian distribution [3]:

1. The mean curvature of the summits is nearly same as the rms value of the peaks.
2. The density of summits (asperities) per unit area is approximately equal to $1.8\eta_p^2$ where η_p is the number of peaks per unit length.
3. The rms value of summit distribution is same as that for general height distribution.

The average curvature of peaks and rms roughness of general height distribution, are usually obtained from the stylus instrument. Hence it is easy to implement the calculations. However while the mean line of general heights is available the mean line for distribution of peaks may not be available and has to be obtained from digitized data of asperity heights. The mean line of the asperities may be displaced upward by about 0.5σ to 1.5σ from the mean line of general heights [3].

In the above formulation each individual asperity contact follows the Hertz theory for elastic contact. Thus for individual asperity the contact area is proportional to $(L_i)^{2/3}$ where L_i is the load acting on the given asperity "i". However when the above equations are solved it can be shown that the total real area of contact is nearly proportional to the load. With increasing load the size of the existing contacts increase and at the same time new smaller contacts form. The rate of formation is such that average asperity size $\bar{a}$ is nearly constant. This leads to the observed proportionality. As Amontons' laws are valid for the case where real area is proportional to load it was concluded that these laws are applicable for elastic contact as well.

The theory can also be used to determine approximately the number of plastic contacts by assuming that an asperity yields plastically when the pressure on the asperity equals $0.6H$.

A simpler criteria proposed to characterize the extent of plastic contact [2] between two surfaces is given by plasticity index ψ defined as

$$\psi = \frac{E}{H}\sqrt{\frac{\sigma}{\beta}} \qquad (1.14)$$

where E, σ, and β are obtained from Eqns (1.11), (1.12), and (1.13).

In the Eqn (1.14) H refers to the hardness of the softer metal.

When the value of the index is >1.0 the contact is considered predominantly plastic whereas it is considered essentially elastic if the value is <0.6.

An example can show the utility of the plasticity index. For the case of two medium carbon steels in contact with $E_{1,2} = 203$ GPa, $H = 1.8$ GPa, $\sigma_{1,2} = 0.2$ µm, and $\beta_{1,2} = 500$ µm. On this basis $\psi = 2.08$.

On the other hand for hard steel surfaces with similar roughness parameters and $H = 7.0$ GPa, the value of ψ calculates to be 0.54. From this criteria the hard steel contact is mainly elastic whereas the contact with softer steel is essentially plastic. The plasticity index may be a good guidance with regard to possible surface damage in the initial run-in stage.

This contact model provides an approximate assessment of elastic contact areas but is restricted to Gaussian distribution. However it may be noted that $p(z)$ is a general distribution function and need not be restricted to Gaussian distribution. In principle if the distribution function is known the contact model can be implemented. Bhushan [4] attempted to solve the problem by treating the roughness as a sum of different mathematical functions. Another approach described here is based on the work of Kumar [5,6] utilizing the bearing area curve. This curve represents the real cumulative distribution of surface heights. It is assumed that the asperity height distribution is similar to general surface height distribution. From this curve distribution function to be used for different zones in contact equations is derived on the basis of the slope of the curve. The coordinates of a point on bearing area curve is obtained from a 2D roughness profile after drawing a horizontal line at a particular height from the lowest valley. The line cuts through solid and empty space. The fractional length that passes through solid can be plotted as a function of height from the lowest valley. This is illustrated in Figure 1.4. The curve

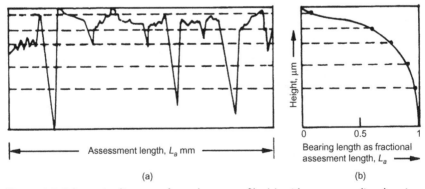

(a)   (b)

Figure 1.4 Schematic diagram of roughness profile (a) with corresponding bearing length curve (b).

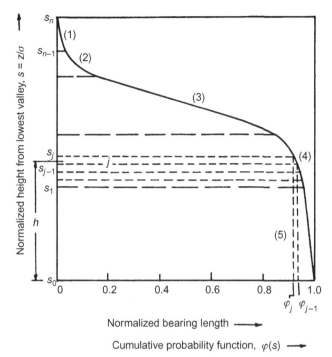

Figure 1.5 Bearing length curve and digitization for non-Gaussian contact model.

will be referred to as bearing length curve though it is conventionally called bearing area curve.

Figure 1.5 depicts the bearing length curve to be used for the non–Gaussian contact model in which x-axis represents fractional length and y-axis represents normalized height, $s = z/\sigma$ from the lowest valley. This curve gives cumulative probability function $\varphi(s)$. Probability density function, $p(s)$, is related to $\varphi(s)$ as $\varphi(s) = \int_h^\infty p(s)\mathrm{d}s$. $p(s)$ may be found by differentiating $\varphi(s)$.

The algorithm to find non–Gaussian contact parameters is as follows:

1. For a load P calculate the value of $F_{1.5}(h)$ using Eqn (1.9).
2. Divide the bearing length curve into different small zones in which variation may be approximated as linear. The slopes for different parts will be the $p(s)$ for that part. By knowing the slopes for different zones, the values of $F_0(h)$, $F_1(h)$, and $F_{1.5}(h)$ mentioned in Eqns (1.7), (1.8), and (1.9) may be found on summing the values of integrals in different zones starting from h to the highest normalized separation S_n (Figure 1.5). The value of h is found numerically by applying the secant method in Eqn (1.17) given below. The $F_0(h)$, $F_1(h)$, and

$F_{1.5}(h)$ may be expressed in terms of normalized height and slopes of different zones as follows:

$$F_0(h) = k_j(s_j - h) + \sum_{i=j+1}^{n} k_i(s_i - s_{i-1}) \qquad (1.15)$$

$$F_1(h) = \frac{1}{2}\left[k_j(s_j - h)^2 + \sum_{i=j+1}^{n} k_i\left\{(s_i - h)^2 - (s_{i-1} - h)^2\right\}\right] \qquad (1.16)$$

$$F_{1.5}(h) = \frac{2}{5}\left[k_j(s_j - h)^{2.5} + \sum_{i=j+1}^{n} k_i\left\{(s_i - h)^{2.5} - (s_{i-1} - h)^{2.5}\right\}\right] \qquad (1.17)$$

where

$n =$ number of zones,

$j =$ the identity of zone in which normalized separation h lies such that $s_{j-1} < h < s_j$ (Figure 1.5),

$k_j =$ slope in jth zone $= (\varphi_{j-1} - \varphi_j)/(s_j - s_{j-1})$,

$\varphi_j =$ normalized bearing length corresponding to normalized height s_j.

3. Record the value of $F_0(h)$ and $F_1(h)$ and then obtain the number of contacts and real area of contact using Eqns (1.7) and (1.8).

4. Finally at a given load, separation, number of contacts, and real area of contact are obtained.

This algorithm is useful for getting contact parameters of any real surface. An example of honed surfaces of an engine liner can show the effect of non-Gaussian surfaces on contact parameters. Consider three types of surfaces—plateau honed, rough honed, and Gaussian—having skewness, S_k of -1.02, -0.46, and 0, respectively. S_k is obtained by taking the third moment of probability density function and it represents degree of symmetry of probability density function. The rms roughness, σ of all the surfaces, was 1.3 μm.

Using the computer program for non-Gaussian contact normalized area of contact, $A_r/(\pi\beta\eta A_n\sigma)$ at normalized load, $P/(E\beta^{0.5}\eta A_n\sigma^{0.5})$ of 0.2 for the preceding three surfaces were found to be 0.23, 0.19, and 0.15, respectively. The real contact area in case of a plateau honed surface is 1.5 times compared to the case of a Gaussian surface. This is a significant difference. Hence non-Gaussian contact analysis will give a more real picture with respect to contact.

Another major problem involved is the scale of observation. For example, the smooth portions observed by a normal stylus instrument may show significant roughness with atomic force microscope (AFM). Similarly we may find that an asperity is made up of smaller scale asperities, which in turn can be resolved into yet smaller asperities. It all depends on the ability to measure at finer and finer scale. Significant research is being conducted now on contact with multiscale asperity distribution. No detailed consideration to this problem is given here. A recent reference [7] along with an earlier paper on the basic formulation of contact [8] is adequate to appreciate the problem. There are contradictions yet to be resolved in these models. It also appears that the models are based on the bulk properties of the materials. This may not be true for very small contact areas.

1.2.2.3 Limitations of the Simple Theory of Friction

The previous explanations for sliding friction are conceptual and qualitative. In the present state of understanding the major limitations of the previous theory may be listed as follows:

1. The simple theory of Bowden and Tabor predicts a friction coefficient of about 0.2 for metals. The values for the dry friction coefficient of metals normally range from 0.2 to 1.0.
2. In elastic contact, though area is proportional to load it is difficult to reconcile with the idea of an adhesive junction formation and their shearing.
3. Even with plastic deformation most of the metals are covered with thin oxides and it becomes difficult to define what the extent of "adhesive" contact is.

In view of the above, alternative explanations are being sought to explain friction and are considered in the next subsection. But before these are introduced, one interesting explanation from the earlier literature [9] regarding growth of contact area shall be considered here.

In the 1950s it was observed that friction in ultra high vacuum between metals with the exception of those with hcp structure keeps increasing. The growth is such that f values greater than 10 can be easily attained and the surfaces can seize over a short distance of sliding. The interesting explanation provided was based on the growth of contact area. The contacts are subject to both normal and tangential stresses and the plastic contact area is obtained based on Eqn (1.18).

$$p^2 + \alpha s_i^2 = \alpha s_m^2 \qquad (1.18)$$

$$f = \sqrt{\frac{\kappa^2}{9(1 - \kappa^2)}} \qquad (1.19)$$

Equation (1.18) represents plastic deformation assuming plane strain conditions. Here p is the normal pressure, α is a constant, s_i is the shear strength of the interface, and s_m is the shear strength of the metal.

From this, the friction coefficient can be obtained as given in Eqn (1.19). Here κ is the ratio s_i/s_m and the value of α is taken as 9.0. These equations show that normal pressure approaches zero as s_i approaches s_m, as is the case for very clean surfaces in an ultra high vacuum. This means for very clean surfaces f values tend to infinity. In many cases the tribo pair just seizes. This approach clearly provides a good explanation at one extreme. Also even small contamination that reduces κ to 0.9 results in an f value of 0.69. This shows strong influence of contamination. In fact many tribological surfaces operate well because they are covered with oxides, reducing interfacial strength. It is possible to arrive at an expression for growth in contact area in terms of the friction coefficient. For relatively low values of f the growth is small.

1.2.2.3.1 Other Aspects of Friction
It may be observed that other factors like ploughing due to hard asperities can also affect friction. Ploughing is important in processes like grinding. Ploughing leading to abrasive wear may also occur due to dust particles as well as hardened wear particles entering the system. Also the treatment here did not distinguish between static and kinetic friction. The value at the onset of sliding is the static friction while the value during sliding is the kinetic friction. Normally kinetic friction is lower than static friction. Depending on the elastic response of the system stick–slip may be observed particularly at low speeds.

Low friction is desirable in many cases from an energy-saving point of view. However there are many cases where friction at a particular level is required; for example, between a shoe and the floor. A dedicated committee of ASTM looks into these problems and devises special test methods. ASTM refers to American Society for Testing and Materials.

1.2.2.4 Models of Macro Friction and Their Status
The friction arises when the asperity contacts are sheared. The earlier models consisted of several extensions and refinements of the basic model of Bowden and Tabor. However the realization that friction can occur with no adhesive junction formation prompted a different kind of adhesion model. Adhesion now refers to the attraction between asperity junctions by van der Waals forces. These forces refer to the physical forces

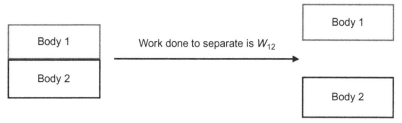

Figure 1.6 Sketch showing separation of two solids.

that are thoroughly researched and understood. These forces result in adhesion that can be treated in terms of surface and interfacial energy. Adhesion as it is normally treated in physical chemistry is illustrated in Figure 1.6.

If two surfaces 1 and 2 are in contact the interfacial energy involved is γ_{12}. The surface energies of the individual surfaces are γ_1 and γ_2. The work of adhesion is defined as the reversible work done, W_{12}, to separate the two surfaces from contact to infinity. The following equation gives the relationships:

$$W_{12} = \gamma_1 + \gamma_2 - \gamma_{12} \tag{1.20}$$

When the two bodies are identical interfacial energy is zero and the work of adhesion is usually called work of cohesion. Usual units used to express surface energy are mJ/m^2 or mN/m.

A conceptual understanding of the problem can be based on a recent thermodynamic model of friction due to Makkonen [10]. In this model the real contact areas that are micro are further subdivided into nano areas. This is like dividing each real area in Figure 1.3 into patches of nano contact. This theory considers work is done in breaking the junctions only and the energy gained in making the junctions does not contribute to the sliding. A thermodynamic argument is provided to justify this approach. The final expression derived when both sliding surfaces are same is

$$f = \frac{\gamma}{d_l H} \tag{1.21}$$

where γ is surface energy of solid, d_l is characteristic dimension of asperity taken as 1 nm, and H is the hardness. The author used surface energy measured by AFM technique for different materials and found good correlation with measured friction in AFM. For example, if a metal surface is involved with γ of $1.0\ J/m^2$ and hardness of 2.0 GPa the f value will be 0.5. The theory is general and in principle applicable to macro contacts as well.

This example is indicative of one approach. Another approach to adhesion is based on adhesion hysteresis [11] and has been applied mainly to lubricant monolayers at nano level. The energy difference between approach and separation in AFM is taken as hysteresis loss. The equation that governs is

$$s = \left(\frac{2\gamma\varepsilon}{d_a}\right) \qquad (1.22)$$

where s is the shear strength, γ is surface energy, ε is the fractional loss due to hysteresis, and d_a is the atomic distance. The hysteresis loss can be 10−20%.

The major problem, particularly at macro level, is to define the surface energy of a solid. Theoretical estimates for metals are difficult and approximate. Also any contamination can vastly change adhesion energy. Metallic materials are covered by thin oxides that can drastically alter energy. For such surfaces there may be no meaningful relation between surface energy and friction. A typical oxide covered steel surface may have low adhesion energy of say $0.1−0.2\,J/m^2$ and Eqn (1.21) may estimate f ranging $0.1−0.2$ while the experimental f may easily exceed 0.5. Such higher friction may be explicable if there is a possibility of primary bond formation in portions of contact and their shearing during the sliding process.

Several other models exist for macro friction. These include adhesion energy coupled with losses due to plastic deformations, models based on dislocations, and possible separation via crack propagation. At the nano level attempts are also being made to model by molecular dynamic simulations. One dislocation model that is interesting [11] argues that asperity contact size can influence dislocations. In effect the shear strength at nano asperities can be very high as dislocations do not propagate. Larger asperities have lower shear strength due to dislocation assisted slip. The detailed modeling of the actual dissipation processes at the atomic level is a major area and is not being considered here. A comprehensive review of nano tribology by Carpick [12] addresses these problems.

The current state of understanding does not help in predicting friction force with any certainty at the macro or nano level. There is significantly more effort at the nano level now due to its future importance in micro and nano technologies. This subsection is intended to provide a brief overview of the issues involved.

1.3 LUBRICATION

Lubrication is usually defined as the separation of moving surfaces by oil or grease, enabling smooth sliding. In a broader sense it may be defined as the complete or partial separation of surfaces by interposed films. Such films reduce friction and wear between the surfaces. This section first considers the important physical properties of a lubricant. It then considers hydrodynamic and EHL that involves complete separation of surfaces. Mixed lubrication and boundary lubrication involving asperity contact are then discussed. Fluid film lubrication is a vast area and the present treatment is limited to an appreciation of the principles. Boundary lubrication will be considered in detail in a later chapter.

1.3.1 Viscosity

When a fluid film is sheared it offers resistance to shear. Consider flow through a channel with the top surface moving with velocity u and the bottom surface stationary. The velocity gradient is du/dz, where z is the height. The resistance to flow, characterized by viscosity η may now be defined as

$$\eta = \tau / \dot{s} \qquad (1.23)$$

where τ is the shear stress, Nm^{-2}, and $\dot{s}$ is shear strain rate du/dz, s^{-1}.

The units of viscosity here are Ns/m^2. The viscosity at a given temperature is constant provided shear stress is proportional to shear strain. This is termed Newtonian behavior. Lubricants normally behave in this manner at moderate conditions, but at high pressures and high shear rates the behavior becomes non-Newtonian.

The viscosity as defined here is called absolute viscosity. Many viscometers measure kinematic viscosity defined as absolute viscosity/density and has units of m^2/s. When expressed in terms of cm^2/s the kinematic viscosity is called stoke. Centistoke, cSt, is the unit commonly used in industry and is equal to 0.01 stoke.

Viscosity varies significantly with temperature. The temperature dependence can be approximated by

$$\eta = A \exp\left(\frac{b^*}{T}\right) \qquad (1.24)$$

where A and b^* are constants and T is the absolute temperature in K.

In practice when viscosity is expressed in cSt Walther equation is used to obtain viscosity at different temperatures accurately. This equation is translated into a convenient chart form and is used in the ASTM D341 procedure. If viscosity at two temperatures is given the line passing through these two points can be used to predict viscosity at any temperature accurately.

Viscosity can change significantly with pressure particularly at high pressures. The basic relation is

$$\eta = \eta_0 \exp(\xi p) \tag{1.25}$$

where

η_0 = absolute viscosity at the given temperature and atmospheric pressure,

ξ = pressure coefficient of viscosity (m^2/N),

η = viscosity at pressure p.

This equation is valid for moderate pressures. Roelands [13] proposed another equation to obtain viscosity considering the influence of temperature and high pressures. This equation has been modified by Hupert and is described by Gohar [14]. This adaption involves defining a new pressure coefficient of viscosity ξ^* for a given condition and using it in place of ξ in Eqn (1.25). Another practical approach is that described by American Gear Manufacturers Association (AGMA) and described in AGMA-925-A03. This is based on experimentally found film thickness in EHL described in Section 1.3.3, and then using the relevant theoretical equation to find ξ.

Another viscosity-related parameter is the Viscosity Index (VI). This is an empirical number to characterize the rate of change of viscosity with temperature. The higher the VI, the lower the variation with temperature. Good lubricants have VI > 100. Detailed methodologies to calculate VI are available in ASTM D2270.

1.3.2 Hydrodynamic Lubrication

The flow of a lubricant through a convergent wedge generates pressure that can support load. This is the mechanism by which a load carrying film is formed between moving surfaces and is referred to as hydrodynamic lubrication. This is illustrated by a slider bearing in Figure 1.7. When the lubricant flows through the convergent wedge, the mass flow rate should be the same. This is possible only through

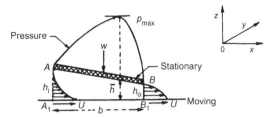

Figure 1.7 Slider bearing showing velocity and pressure distribution.

pressure generation that in turn modifies the velocity distribution across the film. Inlet and outlet velocity distribution is also shown in this figure. Thrust bearings that support axial load utilize this principle. A rotating runner is supported on tapering thrust pads that provide load support.

Hydrodynamic lubrication is governed by the well-known Reynolds equation for two-dimensional flow:

$$\frac{\partial}{\partial x}\left(h^3 \frac{\partial p}{\partial x}\right) + \frac{\partial}{\partial y}\left(h^3 \frac{\partial p}{\partial y}\right) = 12\bar{u}\eta_0 \frac{\partial h}{\partial x} \qquad (1.26)$$

where h = film thickness at (x, y) and $\bar{u} = (u_1 + u_2)/2$, where u_1 and u_2 are the sliding velocities of the two surfaces.

This equation is based on incompressible flow and the assumption there is no flow in the z direction. Flow in the z direction occurs in some cases like dynamically loaded bearings, in which case the 3D version of Reynolds equation is applied. The flow of the lubricant is in x and y directions, though generally the flow in the y direction is smaller. The flow in the y direction is also called side leakage. The equation can be solved through finite difference methods. Graphical procedures of Raimondi and Boyd [15,16] are also used extensively to solve the 2D problem.

In the case of journal bearing that supports radial load the same theory is applicable. However in this case convergent and divergent film is involved and selection of boundary conditions presents some difficulty. Hamrock [17] has given a detailed coverage of all aspects of hydrodynamic lubrication.

1.3.3 Elastohydrodynamic Lubrication

This regime of lubrication occurs in concentrated line and point contacts. Such contacts occur in components like gears and rolling

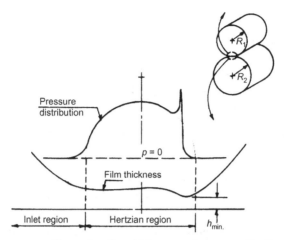

Figure 1.8 Line contact illustrating film thickness and pressure distribution.

element bearings and the stresses involved are high. The stresses and their distribution are governed by Hertz theory and are covered elsewhere in this book. When high stresses are involved the elastic deformation and pressure effect on viscosity become significant. The Reynolds equation now has to be solved with due consideration of these effects by numerical methods. The early developments for line and point contacts were due to Dowson, Higginson, and Hamrock [18,19]. The nature of EHL for line contact is illustrated in Figure 1.8. In this figure p refers to gage pressure.

The figure illustrates an important difference from normal hydrodynamic lubrication. Most of the film is parallel except for a constriction toward the exit. The pressure peak exists before the position of minimum film thickness. The relation that governs film thickness is

$$\frac{h_{\min}}{R} = 2.65 \frac{G^{0.54} U^{0.7}}{W^{0.13}} \tag{1.27}$$

where

$h_{\min}$ = minimum film thickness (m),

$G = \xi E^*$, dimensionless material parameter,

$U = \eta_0 \frac{u_1 + u_2}{2E^* R}$, dimensionless speed parameter,

$W = \frac{w}{E^* R L_x}$, dimensionless load parameter,

$R = \left(\frac{1}{R_1} + \frac{1}{R_2}\right)^{-1}$, m,

$$E^* = \left(\frac{1}{2} \left(\frac{1-\nu_1^2}{E_1} + \frac{1-\nu_2^2}{E_2} \right) \right)^{-1}, \mathrm{N/m^2},$$

w = total load in the contact (N),

L_c = length of contact (m),

u_1, u_2 = velocities of the two surfaces (m/s).

Similar equations with some change in exponents are also available for point contacts. These equations are for the E-V condition that is normally prevalent. This refers to a situation where both elasticity (E) and variable viscosity (V) influence film thickness. There can be situations where only one factor may be dominant. For example, if one of the rollers is made of rubber, elastic deformation will be significant but pressure effect on viscosity will be negligible. Equations are now available for each specific condition. The present approaches and several examples of film thickness calculations are available in the book by Szeri [20]. Film thickness in EHL contacts ranges from 0.1 to 1.0 μm and are an order of magnitude lower than in hydrodynamic lubrication. Also from this equation it can be seen that influence of load is very small on film thickness. It may be noted that reduced modulus of elasticity is defined differently when compared to Eqn (1.11).

One important issue in EHL is related to transient conditions. This equation is for steady state and suitable adaptation is needed for real situations that occur in gears and other concentrated contacts where sliding speeds and stresses can change along the contact line. Earlier solutions were approximate. Present research is focused on transient conditions and mapping of film thickness accurately. A detailed consideration is available in the 30th Leeds-Lyon symposium [21].

1.3.4 Thinning Films and Boundary Lubrication

It is desirable to operate tribological contacts with full separation of surfaces by fluid film. This occurs when minimum film thickness is greater than 3σ, where σ is the composite roughness defined by Eqn (1.13). The ratio of minimum film thickness to composite roughness is called lambda ratio and is defined as

$$\Lambda = \left(\frac{h_{\min}}{\sigma} \right) \tag{1.28}$$

Several components operate with Λ values less than 3.0 leading to asperity contact. Some examples include cam-tappet and ring-liner contacts in IC engines as well as gears with high sliding components like worm and hypoid gears. As asperity contact occurs, the load is shared between asperity and fluid film; this zone is called the mixed lubrication zone. This zone is considered to occur until a Λ value of about 0.5. Below this value the load is mainly supported by asperities via molecular films adsorbed on the surface. This zone is called boundary lubrication and is the main zone of interest in this book. It needs to be emphasized that in mixed lubrication asperity contact really involves contact via boundary films and is not direct metallic contact. Mixed lubrication is complex because film thickness is also affected by roughness for thin films. Several interesting approaches are being tried to reduce metal contact in mixed lubrication. One approach that has been successful is the laser texturing of surfaces [22,23]. This texturing helps in increasing the hydrodynamic effect as the micro dents act as miniature step bearings.

NOMENCLATURE

$\bar{a}$	average asperity size
A	constant
A_n	nominal (geometric) contact area
A_r	total real contact area
b^*	constant
d	apparent contact diameter
d	separation between rigid plane and the mean line
d_a	atomic distance
d_l	characteristic dimension of asperity
E	effective modulus of elasticity
E^*	effective elastic modulus defined as $2E$
E_1, E_2	elastic modulus of the two bodies in contact
f	coefficient of friction
F	friction force
$F_n(h)$	$= \int_h^\infty (s-h)^n p(s) ds$
G	dimensionless material parameter
h	d/σ standardized separation between rigid plane and the mean line
h	film thickness at (x,y) (Eqn (1.26))
$\bar{h}$	film thickness at $dp/dx = 0$
h_i, h_o	film thickness at inlet and outlet in slider bearing
h_{min}	minimum film thickness
H	hardness of the softer material
k_j	slope in jth zone of bearing length curve
L	load in dry contact
L_a	assessment length of roughness profile
L_c	length of EHL line contact
L_i	load acting on the asperity i
n	number of asperity contacts

n	number of zones in bearing length curve in which slope is assumed to be constant
p	pressure
$p(z)$	asperity height distribution function
$p(s)$	standardized asperity height probability density function
p_{max}	maximum pressure
P	load
R	equivalent radius of cylinder
R_1, R_2	radii of cylinders 1 and 2
R_a	center line average surface roughness
R_q	the root mean square (rms) surface roughness
R_z	the mean distance between the five highest peaks and five lowest valleys within the sampling length
s	shear strength
s	standardized asperity height over mean line
$\dot{s}$	shear strain rate, du/dz
s_i	shear strength of the interface
s_m	shear strength of the metal
S_k	skewness, measure of asymmetry of the surface height distribution, third moment of surface height from the mean line
T	absolute temperature
$\bar{u}$	mean surface velocity defined as $(u_1 + u_2)/2$
u	sliding velocity of top surface with respect to stationary bottom in a channel flow
u_1, u_2	the sliding velocity of two surfaces in lubricated contact
U	dimensionless speed parameter
w	total load in EHL contact
w	load support in thrust bearing
W	dimensionless load parameter
W_{12}	work of adhesion, reversible work done to separate the two surfaces 1 and 2 from contact to infinity
z	asperity height over mean line
z	distance between moving top surface with velocity u from the stationary bottom surface in a channel flow
z_i	ith asperity height over mean line

Greek Letters

α	constant
β	$= \beta_1\beta_2/(\beta_1 + \beta_2)$ average asperity radius of curvature of an equivalent surface
β_1, β_2	average asperity radius of curvature of two surfaces in contact
δ_i	contact deformation of asperity $i = (z_i - d)$
ε	fractional loss due to adhesion hysteresis
γ	surface energy of solid
γ_1, γ_2	surface energies of individual surfaces 1 and 2
γ_{12}	interfacial energy if two surfaces 1 and 2 are in contact
η	number of asperities per unit area
η	absolute viscosity
η_0	absolute viscosity at the given temperature and atmospheric pressure

η_p number of peaks per unit length

$\varphi(s)$ cumulative probability density function

φ_j normalized bearing length corresponding to normalized height s_j

κ ratio of shear strength of the interface to the shear strength of the metal, s_i/s_m

Λ ratio of film thickness to roughness

ν_1, ν_2 Poisson's ratios of two bodies in contact

σ_y yield strength

σ $= \sqrt{\sigma_1^2 + \sigma_2^2}$ rms roughness of the equivalent surface

σ_1, σ_2 rms roughness of surfaces 1 and 2

τ shear stress

ξ pressure coefficient of viscosity at the given temperature

ξ^* new pressure coefficient of viscosity for a given condition

ψ plasticity index

REFERENCES

[1] Bowden FP, Tabor D. The friction and lubrication of solids, part II. London: Oxford University Press; 1964. [Chapter IV].

[2] Greenwood JA, Williamson JPB. Contact of nominally flat surfaces. Proc R Soc Lond A 1966;295:300.

[3] Johnson KL. Contact mechanics. London: Cambridge University Press; 1985. p. 410.

[4] Kotwal CA, Bhushan B. Contact analysis of non-Gaussian surfaces for minimum static and kinetic friction and wear. Trib Trans 1996;39(4):890–8.

[5] Kumar R, Prakash B, Sethuramiah A. Theoretical modeling of surface contact for honed surfaces. In: Bhatia J, editor. Advances in industrial tribology. New Delhi: Tata McGraw Hill; 1998. pp. 100–7.

[6] Kumar R, Investigation into the running-in and steady state wear processes [Ph.D. thesis]. Delhi: IIT; 2002.

[7] Jackson RL, Streator JL. A multi-scale model for contact between rough surfaces. Wear 2006;261:1337–47.

[8] Adams GG, Nosonovsky M. Contact modeling—forces. Trib Intl 2000;33:431–42.

[9] Tabor D. Junction growth in metallic friction: the role of combined stresses and surface contamination. Proc R Soc Lond A 1959;251:378.

[10] Makkonen L. A thermodynamic model of sliding friction. AIP Adv 2012;2 (012179):1–9.

[11] Hurtado JA, Kim KS. Scale effects in friction of single−asperity contacts. I. From concurrent slip to single-dislocation-assisted slip. Proc R Soc Lond A 1999;455:3363–84.

[12] Szlufarska I, Chandross M, Carpick RW. Recent advances in single−asperity nanotribology. J Phys D Appl Phys 2008;41:123001, 1−39.

[13] Roelands CJA. Correlation aspects of the viscosity-temperature-pressure relationship of lubricating oils. Groningen: Druk, V. R. B.; 1966.

[14] Gohar R. Elastohydrodynamics. Chichester: Ellis Horwood Limited; 1988.

[15] Raimondi AA, Boyd J. Applying bearing theory to the analysis and design of pad−type bearings. ASME Trans 1955;77(3):287.

[16] Raimondi AA, Boyd J. A solution for the finite journal bearing and its application to analysis and design (in three parts). ASLE Trans 1958;1:159.

[17] Hamrock BJ, Schmid SR, Jacobson BO. Fundamentals of fluid film lubrication. 2nd ed. New York, NY: Marcel Dekker, Inc.; 2004.

[18] Dowson D, Higginson GR. Elastohydrodynamic lubrication. Oxford: Pergamon; 1977.

[19] Hamrock BJ, Dowson D. Ball bearing lubrication—the elastohydrodynamics of elliptical contacts. New York, NY: Wiley; 1981.

[20] Szeri AZ. Fluid film lubrication: theory and design. Cambridge: Cambridge University Press; 1998.

[21] Dalmaz G, Lubrecht AA, Dowson D, Priest M, editors. Transient processes in tribology: proceedings of the 30 Leeds-Lyon symposium on tribology, INSA de Lyon Villeurbanne, France, 2—8; September, 2003.

[22] Brizmer V, Kligerman Y, Etsion I. A laser surface textured parallel thrust bearing. Trib Trans 2003;46(3):397—403.

[23] Kovalchenko A, Ajayi O, Erdemier A, Fenske G, Etsion I. The effect of laser texturing of steel surfaces and speed-load parameters on the transition of lubrication regime from boundary to hydrodynamic. Trib Trans 2004;47(2):299—307.

CHAPTER 2

Lubricants and Their Formulation

2.1 INTRODUCTION

A lubricant may be defined as a solid or liquid film that is interposed between surfaces to reduce friction and wear as stated in the previous chapter. A lubricant has many other functions besides lubrication alone. In all cases it removes the heat from the surfaces and carries away wear debris that gets removed through filters. It serves additional purposes mainly by the incorporation of additives that are system specific.

This chapter first deals with the manufacture and refining of petroleum-based oils called base oils. Common additives and the formulation technology form the next section. This is followed by a discussion of the lubricant specifications and the test methods involved. The next section deals with synthetic lubricants. The final two sections deal with the issues related to environment and performance. Several books are available on lubricants and the present treatment is limited to an appreciation of the main aspects of lubricant technology.

2.2 PRODUCTION OF BASE OILS

2.2.1 Crude Distillation

Crude petroleum as received is first subjected to distillation to remove the lighter ends as side streams of varying boiling ranges. The lighter components consist of gases, gasoline/naphtha, kerosene, and diesel cuts. The heavier side stream is called gas oil, which may be converted to lighter products by cracking. The bottom residue called reduced crude contains the high molecular weight components suitable for manufacture of the lubricants.

The reduced crude is then subjected to vacuum distillation to obtain base oils of different viscosity ranges. Base oils are also called base stocks. Figure 2.1 shows the basic distillation processes involved. Spindle oil is the lightest and heavy neutral is the heaviest stream. The numbers associated with neutral refer to the older Saybolt viscosity numbers. The vacuum residue in some cases can be deasphalted to produce a very high viscosity oil called bright stock. The viscosities measured at 40°C range from 10 cSt for spindle oil to as high as 600 cSt for bright stock.

Modeling of Chemical Wear.
DOI: http://dx.doi.org/10.1016/B978-0-12-804533-6.00002-0
© 2016 Elsevier Inc.
All rights reserved.

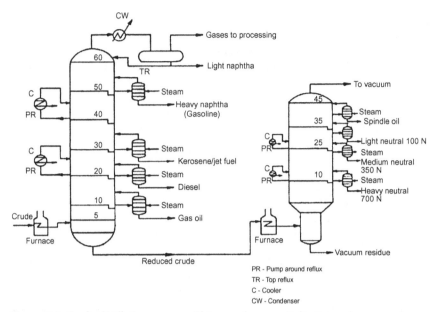

Figure 2.1 Crude distillation process. Plate numbers are indicative only.

Note that only some crudes that contain required high molecular weight components are suitable to obtain base oils. The base stocks produced are to be further refined to use them as lubricants. Also the words lubes, lubricating oils, base oils, and base stocks are all used in lubricant technology and have to be understood in the context used. For the purpose of this chapter base oil or base stock is used for the raw oil fractions obtained from refinery. These stocks after refining are called finished base oil or finished base stock. Lubricating oil is the term used for the final blended product to be used for specific applications.

2.2.2 Refining of Base Stocks

The refining of base stocks is to remove undesirable components. The undesirable components can be removed by physical processes like solvent extraction or by converting the components into desirable components by chemical processes. The processes used depend on the quality of base stocks.

The major base stocks used are paraffinic. Naphthenic oils are produced in limited quantities and are well suited for low temperature applications. Paraffinic oils have straight chain and isomeric alkanes that have desirable properties like high viscosity index and oxidation stability.

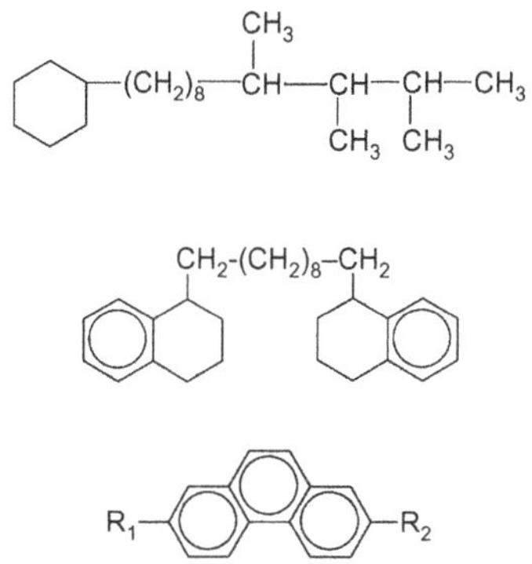

(a) Long isoparaffin chain linked to naphthenic ring.

(b) Two naphtheno-aromatic rings bridged by normal paraffin chain.

(c) 3-Ring condensed aromatic linked to two small alkyl groups.

Figure 2.2 Typical hydrocarbon structures in paraffin base oil.

But it may be emphasized that most of the hydrocarbons have mixed structure with naphthenic or aromatic rings. Paraffinic oils only mean that the alkane part of the structure predominates. The base oils range in molecular weight from C_{25} to C_{40}. The oils also contain a small percentage of predominantly aromatic or naphthenic components. The diagram in Figure 2.2 shows some typical structures found in base oils. The paraffin or isoparaffin structures can be linked to naphthenic or aromatic structure. One essentially aromatic structure that is undesirable is also illustrated.

The base oil needs upgrading in three areas. One is the removal of aromatic components to improve VI and oxidation stability. Second is the removal of very long-chain normal paraffins (dewaxing) to improve low temperature flow properties. The third is finishing processes to remove impurities like sulfur and oxygen. The processes involved are as follows:

- **Aromatic removal** is achieved by selective solvent refining with solvents like furfural. This involves some loss of product. The other process is hydrogenation, which saturates the aromatic chains. There is no loss in this case and the process is widely followed now.
- **Dewaxing** is done in a conventional process with methyl ethyl ketone as solvent, chilling and precipitating the wax, and final filtration. Modern processes are catalytic, in which the long-chain paraffins preferentially pass through a zeolite bed and undergo mild cracking and later are subjected to isomerization.
- **Finishing** involves a mild hydrogenation process that eliminates impurities by reaction. For example, sulfur can get removed as hydrogen sulfide. Color is also improved in the finishing process.

One additional process that is gaining major importance is hydrocracking. Originally envisaged for cracking heavy products to fuels, it now has major flexibility and is capable of combining cracking and hydrogenation to the required level. Thus saturation of aromatics, isomerization, and opening of naphthenic rings leading to paraffins are all possible in this process. The flexibility is such that some feeds like aromatics can be converted to desirable lubricant molecules.

Finished oils are required to meet specified physicochemical properties that are ensured by the refiner. These tests include viscosity, VI, laboratory distillation, sulfur content, and aromatic content, besides others. ASTM has a large number of tests for this purpose [1]. Modern petroleum technology [2] covers all the processes discussed here in detail.

2.3 ADDITIVES AND FORMULATION TECHNOLOGY

2.3.1 Nature of Additives

Finished base oils obtained from the refinery is the starting point to manufacture lubricants for the industry. The oils are first blended to obtain the desired viscosity grade. Then depending on the application additives are added to enhance certain properties. The final lubricant has to meet the specification for that particular oil. For example turbine oil needs an oxidation stability level that cannot be met by finished base oil alone. Antioxidant additives are added to bring up the oxidation stability to the desired level. Another example is gear oil used for severe conditions that is blended with chemical additives to protect against scuffing. In some cases a large number of additives are to be incorporated as is the case with engine oils. The major classes of additive used and their purpose are given in Table 2.1.

Besides these classes there are other additives like pour point depressants that are less used now in view of the superior quality of base oils. It may be also noted that emulsifiers are common additives for metal working fluids to stabilize oil in water emulsions. Detailed coverage of lubricant additives is available in the book edited by L. R. Rudnick [3].

The formulation is obtained by blending the additives in the required dosage with the selected base stock. The additives used may be "packages" obtained from additive manufacturers or can be special formulations of the blending oil company. The formulations are tested against specifications and then marketed. For known standard formulations the blending is routine and is carried out in modern plants. Initial development of formulations is an intricate task where several issues are to be taken into account.

Table 2.1 Major additive classes used in lubricants

Additive	Typical chemical types	Purpose
Detergent	Sulfonates, phenates	To keep carbon in suspension in engine oils
VI improver	Olefin copolymers, polyalkyl methacrylates	To provide higher VI
Dispersant	Succinimides, succinate esters	To keep sludge in suspension in engine oils
Antioxidants	Phenolics and amines	Protect base oil from oxidation
Antiwear	Less reactive sulfur and phosphorous compounds	Reduce wear by forming wear-resistant reaction films on surfaces
Extreme pressure (EP)	More reactive sulfur compounds	Prevent scuffing by reacting fast and provide protection
Multifunctional	Zinc dithiophosphates	Functions as antiwear and antioxidant; invariably used in engine oils
Friction modifier	Long-chain fatty acids and esters	Form thin films on the surfaces that shear easily

2.3.2 Development of Formulations

The major issues involved in new formulation are the additive interactions, the changing needs of the equipment manufacturers, and specifications. Environmental issues and their impact are discussed elsewhere in this chapter. These issues are interconnected but considered here separately. The major specifications are considered in the next subsection. The situation is dynamic and constantly evolving. The first issue is whether the additive combination is compatible.

2.3.2.1 Additive Interactions

Interactions are given in Figure 2.3. The desirable interaction is synergism or at least compatibility. Antisynergism is to be ruled out in formulations. Another aspect is the need for compatibility of different marketed products. Such a need exists, for example, for engine oils. A motorist may top up the oil from any company and there should be compatibility of the oils. Such compatibility is ensured by the oil companies.

The information regarding additive compatibility is well known to the oil industry and a good knowledge base regarding common additives is available. The problems arise when additives with new chemistry are involved.

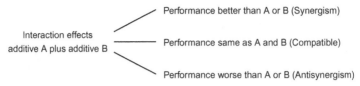

Figure 2.3 Additive interactions.

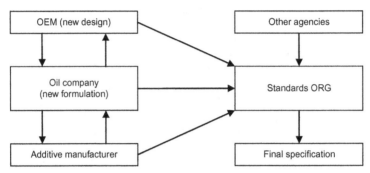

Figure 2.4 Development of specifications.

2.3.2.2 Equipment Upgradation

The major customers of the oil industry are known as original equipment manufacturers (OEMs). The OEMs keep upgrading their equipment and want the lubricants to work effectively for the modified systems. Two kinds of problem arise. One is the inadequacy of the existing oil that has to be improved. The second is the creation of specifications with the required tests to make sure the lubricant is adequate. The evolution of standard engine tests is a case in point.

Essentially the modern designs that are compact and efficient result in the need for lubricants that withstand increased severity. Such improvements result in periodic changes in quality. Such changes may call for new or modified specifications. These changes translate into more stringent requirements on the specified laboratory tests including mechanical rig tests. In some cases completely new tests may be developed and incorporated. While this is most evident for internal combustion (IC) engines it is also applicable to other components like highly loaded gears, cold rolling at very high speeds, and high-speed cutting. Besides OEMs and the oil industry many other agencies that have a stake in standards are also involved in creating specifications. Figure 2.4 depicts the situation.

2.4 LUBRICANT SPECIFICATIONS

First, for every category of the lubricant physicochemical properties of relevance are given with appropriate limits. The specifications also include laboratory rig tests along with expected performance. There may be multiple specifications developed by defense, national standard organizations, private institutions, and international organizations. In many cases they draw on each other's experience. It is common to divide lubricants into automotive and industrial categories. Automotive lubricants consist of engine and gear oils. Other lubricants like automatic transmission fluids and greases are not considered here. Engine oils form about half of all the lubricants manufactured and thus deserve to be treated as a special category.

2.4.1 Automotive Lubricants

Engine oils are the major category of automobile lubricants. Engine oils are divided into grades based on viscosity. The widely used classification is based on Society of Automotive Engineers (SAE) specification. The latest version as of this writing is J-300, 2013. Engine oils are divided into winter and summer grades. The summer grades range from 20 to 60 with viscosity at 100°C ranging from 5.6 to 26.1 cSt. The winter grades range in viscosity from 3.8 to 9.3 cSt at 100°C. The winter grades range from 0 to 25 W. For winter grades special tests for low temperature cranking and pumping are also specified. Also a minimum for high shear viscosity at 150°C is specified. High shear viscosity is important with regard to high shear zones like bearings. A minimum viscosity ensures reduced asperity contact.

The performance with regard to detergency, sludge, corrosion, and wear are determined by special engine tests. These tests are necessary as simpler laboratory tests are unsatisfactory for this purpose. The specifications will have several categories for gasoline and diesel engines. The categorization is based on the severity of operation. Each category will have its specific engine tests to be conducted. For example following the classification by the American Petroleum Institute (API) there are four categories for diesel engine oils. The oil for heavy duty is designated CJ-4 and has eight different engine tests specified to determine several aspects of performance including oil thickening, soot, detergency, cam-tappet wear, and ring wear. Each engine test follows standardized procedure as per ASTM. The details of this major area of testing are not within the scope of this book. The details related to API will be available at their site, www.api.org/eolcs. A recent book covers modern details of automotive lubricants [4].

Operational severity today in engines is significantly higher than a decade ago. This in turn calls for new formulations and new tests. The better quality lubricants now contain more stable high VI oils. There are five types now covered by API. Types I, II, and III are petroleum based while IV and V are synthetics. Type III has a VI of >120 and is well suited for engine oils.

The viscosity grades for gear oils are governed by SAE J306. These cover nine grades ranging from 70 W to 250 with corresponding viscosity ranges of 4.0–41 cSt at 100°C for gear boxes. For the hypoid rear axles special performance tests are called for as these axles are prone to scuffing due to high levels of sliding. Automatic transmission fluids have special needs like freedom from shudder and good control over friction. The tests developed by US manufacturers GM and Ford are popularly used for this purpose.

2.4.2 Industrial Lubricants

Major classes of industrial lubricants are metal working fluids, hydraulic oils, turbine oils, and gear oils. The products under each category are vast in number. For example, metal removal processes involve cutting, drilling, tapping, honing, grinding, and milling. Each operation again has many variations—cutting may involve large variations in tools from high carbon steel to ceramics. For ceramics lubricants cannot be used. Also cutting rates can vary over a broad spectrum. So specifications become difficult and generally rely on minimum performance in a laboratory machine. These are tribological tests. Thus real performance has to be based on experience with actual products that compete for the market. Several books deal with these lubricants and test methods [5,6]. The main requirements are covered in Table 2.2.

2.5 SYNTHETIC LUBRICANTS

Synthetic lubricants may be defined as those fluids that are chemically synthesized, unlike mineral oils obtained from crudes. Unlike mineral oils the molecular size of many synthetics can be effectively controlled. This provides a great advantage in selecting the right fluid with controlled properties. Another advantage is the chemical purity of the fluid. This is unlike the mineral oils that may have a small percentage of undesirable structures that may affect, in particular, the high temperature

Table 2.2 Industrial lubricant categories

Lubricant type	Major requirements
1. Metal working a. Oil in water emulsions b. Neat cutting oils for heavy duty c. Metal forming lubricants-emulsions and sometimes synthetics	a. emulsion stability and antifungal properties b. EP additives to control pick-up and scuffing tested as per ASTM D2266 and D3233. c. Emulsion with effective oil separation in roll bite coupled with need for boundary contact. In-house tests and experience based. Synthetics for very high speed rolling.
2. Hydraulic oils	Mild chemical additive for wear control coupled with good oxidation stability. Vane pump tests to find wear. Oxidation tested by ASTM D943.
3. Turbine oils a. R&O type b. EP type	a. High level of oxidation stability and rust inhibition required. Tested as per standard procedures like D943. b. This is for cases where reduction gears are involved. EP protection tested in FZG rig as per DIN 5134.
4. Industrial gear oils normal category and EP category	Usually covered under AGMA specifications that have nine categories based on viscosity at 40°C. EP levels to be checked by FZG rig. Oxidation stability, rust, foaming requirements also specified. Other specifications are due to US Steel and David Brown.

FZG rig; German standard gear test rig.

stability. But it may be noted that highly refined stocks based on hydrogenation are now competing well with synthetic base stocks and the scenarios are changing. This section will consider the types of synthetics and their applications.

2.5.1 Types of Synthetic Lubricants and Their Applications

There are a very large number of synthetic fluids and the present consideration will be limited to commonly used lubricants. The major classes and their structures are given in Figure 2.5. Several other products like

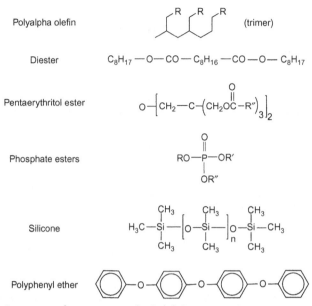

Figure 2.5 Structures of common synthetic lubricants.

silicate esters, silahydrocarbons, and chlorofluoro polymers that are used for very special applications are not covered here.

Synthetic lubricants have both desirable and undesirable properties. They have to be used with great care. Usually they are for applications where mineral oils cannot perform because they are generally expensive in comparison to mineral oils. Also within each class there can be large variations in properties. Newer molecules with still better properties are continuously being explored.

Polyalphaolefins (PAOs) and diesters are finding increased usage in engine oils, particularly in Europe. They have superior VI, excellent low temperature fluidity, and thermal and oxidative stability. They also show superior high temperature, high shear rate viscosity. The esters are easily biodegradable. The formulations can be fully synthetic or mixtures of mineral and synthetic fluids. PAOs are also used for aircraft hydraulic systems. At present very high VI oils of 120 and above (API type III) are being manufactured by hydroprocessing as stated earlier. Such base stocks have become competitive to PAOs and are cheaper. They are finding more use in engine oils.

Several combinations of diesters and monoesters like pentaerythritol esters are used for aviation gas turbine lubrication. In fact major developments in synthetics are due to civil and military aviation.

The diesters are poorer with regard to hydrolytic stability. They can also attack normal paint materials and seals. Due care in these areas is necessary for using esters.

Phosphate ester fluids are unique for their fire resistance; their original applications were as aircraft hydraulic systems. They also have good lubricity. Their main problem is hydrolytic stability since they easily react with water generating phosphoric acid. They also find application in compressors and brake fluids.

Polyphenylethers have excellent thermal and oxidative stability but have poor low temperature flow properties and low VI. Their use is now minor in aviation. They are used in the nuclear industry due to their radiation resistance. They also find use as lubricants in some satellite applications where their low evaporation loss is important.

Silicon containing fluids are silicones, silicate esters, and silahydrocarbons. Silicones are the more commonly known fluids. They have excellent stability, high VI, and low temperature fluidity. However they are very deficient in lubricity. They find limited application in vacuum greases. Silicate esters have problems of hydrolytic stability and are less used. Silahydrocarbons are fluids that reduce the disadvantages of other fluids but their use is still experimental.

Perfluoropolyethers are very expensive lubricants and at present have an interesting application. They are used as a lubricant for computer hard disks. The application involves formation of self-assembled monolayers on the surface. Their main purpose is to prevent stiction. Stiction is caused in these systems at the disc—head interface due to water meniscus formation. This meniscus formation is prevented by the monolayer.

2.6 ENVIRONMENTAL ISSUES

Environmental issues are of major concern today. Several reports at the national and international levels have brought a major awareness about the dangers to mankind if we move recklessly forward. For example global greenhouse gases have increased by 70% during 1970—2004 and are responsible for global warming, leading to climatic changes. It is important to know how the main stream lubricants and additives affect the environment. The remedial measures that can be adopted are then considered.

2.6.1 Effect of Lubricants on Environment

First, we must consider the extent of lubricant contribution to greenhouse gases. The contribution is mainly from IC engine exhaust. The ratio of fuel burnt to lubricant is estimated as 410:1. It is hence considered that this pollution is not majorly due to lubricants. However lube additives that go into the atmosphere via burnt lubricant are significant. It is estimated that 50% of particulate emissions from engines are due to lubricant additives.

The second issue is disposal of used lubricant. It is difficult to estimate the various ways of disposal. The data based on the report by Technical Committee of Petroleum Additive Manufacturers of Europe [7] is considered here. This report is limited to crank case oils. The mass balance for the oil is as follows:

- Exhaust: 24%
- Incineration with energy recovery: 36%
- Regeneration (Rerefining): 18%
- Irregular disposal: 21%

The irregularly disposed oils are a major worry because these oils pollute water bodies and soil. With regard to industrial lubricants as much as 70−80% of the drained oils end up as pollutants. The statistics may be similar for many countries outside Europe.

The negative environmental effects are due to the base oils as well as additives. The major effects are on aquatic life. These effects are toxicity on fish, bacteria, and mammals, and standard tests are available under German standard DIN 38 412-15 and other procedures due to ASTM. Besides toxicity, biodegradability of the used lubricants is also important. A series of methods for this are available through the test methods of Organization of Economic Cooperation and Development numbering 301 to 312.

2.6.2 Approaches for Improvement

2.6.2.1 Modifying the Existing Products

The existing base stocks are mostly petroleum-based and have poor biodegradability. As discussed earlier in the chapter synthetics like diesters are more biodegradable and are also miscible with mineral oils. Hence the use of diesters in crank case oils is a right step in this direction. Several lubricants with diester base or mixed with mineral oil are already on the market.

Changing the additives is a more complex problem since performance is well controlled by them. The current efforts are aimed at developing ashless additives both for detergents and zinc dithiophosphate used in crank case oils. These developments are in the research stage now.

2.6.2.2 New Eco-Friendly Formulations

Another and better approach is to encourage use of bio-based materials like vegetable oils and their modifications. Such products are being increasingly used for moderate duty applications. These applications include hydraulic fluids for farm equipment, cutting fluids, chain saw lubricants, and other miscellaneous uses. These applications in farming and forestry applications are encouraged by governments. This is especially true with Germany and Nordic countries that have a keen interest. Subsidizing the costs incurred is one of the steps to encourage biolubricants. Complete specifications are drawn up for these products by various agencies. In the United States the farm bill and the US federal bio-based product preferred procurement program encourage biolubricants.

Some nongovernmental agencies are actively involved in the promotion. For example the eco label "Blue Angel" in Germany is a well-known recognition. Some other labels include Nordic Swan and Canadian Eco logo.

Application of these stocks for heavy duty is difficult. One very serious limitation is the oxidation stability at high temperatures. This area is under investigation but it appears the products will be confined to medium-duty applications in the near future.

2.7 THE ART AND SCIENCE OF PERFORMANCE

Throughout this chapter it has been stated that selection can be based on meeting the prescribed specification. This may be partly true for engine oils but for the highly varied industrial oils including biolubricants the tests are at best indicative. The acceptance values are based on experience and may not be valid for new formulations. So performance tests may be viewed with some caution. A recent paper on the development of vegetable-based hydraulic oil is a case in point [8]. The authors tried to correlate the behavior of different types of vegetable oils in the standard ASTM D 943 dry test for oxidation and the wear behavior in actual vane pumps. It was found that when phenolic and amine anti-oxidants were used, the deterioration process itself is different in the two tests. Thus in pump tests the deterioration is more due to thermal degradation than oxidation.

Such examples abound in literature. Poor correlation between laboratory test and the real system with regard to wear tests were also reported in earlier literature [9]. Real formulations need more insight than just going by test values. There may hence be scope for calling it the art of formulation. The advantage is that a lot of experience is built around successful formulations. Such base is useful in developing new formulations that are modifications. The formulation challenges come when completely new fluids and additives are involved. New formulations need to build up the database again and it means time. Procedures that map behavior over a range of conditions may be more useful in assessment than a go/no-go type of test. Also where possible the data should be analyzed scientifically. An example can illustrate this. While dealing with turbine oil oxidation one of the authors faced a question. If three marketed turbine oils with same additive chemistry are all meeting specifications, which one should be chosen purely from a technical point of view? The attempt done was to first obtain rate constant k at different temperatures in a rotary bomb oxidation tester assuming pseudo first-order reaction. Then an Arrhenius plot of ln(k) versus temperature was drawn. The activation energy obtained was used for performance comparison. This effort was due to C. R. Jagga and A. Sethuramiah and presented in a workshop [10]. An example of a final plot for one turbine oil A is given in Figure 2.6.

The purpose of this plot is to show that analysis over a range of conditions is useful. It is to illustrate a different approach and is not

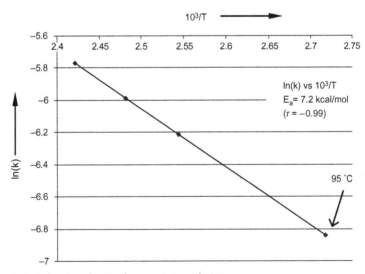

Figure 2.6 Arrhenius plot to characterize oxidation.

claimed as the best approach. It is considered that rethinking about laboratory test methodologies is necessary for effective product evaluation.

Tribological tests are more problematic and are not taken up here. They will be considered in detail in later chapters.

NOMENCLATURE

k   rate constant, min^{-1}
E_a activation energy, kcal/mol
T   temperature, K
r   correlation coefficient

REFERENCES

[1] Methods of tests for petroleum products and lubricants, vols. 5.01 to 5.04. Am Soc Test Mater ASTM International, West Conshohocken, PA 1998.
[2] Meyers RA, editor. Handbook of modern petroleum technology. 2nd ed. New York: McGraw-Hill; 1997.
[3] Rudnick LR, editor. Lubricant additives: chemistry and applications. 2nd ed. Boca Raton, Florida: CRC Press; 2009.
[4] Tung SC, Totten GE, editors. Automotive lubricants and testing. Warrendale PA: SAE International; 2012.
[5] Mortier RM, Orszulk ST, editors. Chemistry and technology of lubricants. Glasgow: Blackie; 1992.
[6] Nachtman ES, Kalipakjian S, editors. Lubricants and lubrication in metalworking operations. New York: Marcel Dekker; 1985.
[7] ATC Document 49 (revision 1), Lubricant Additives and The Environment; 2007.
[8] Petlyuk AM, Adams RJ. Oxidation stability and tribological behavior of vegetable oil hydraulic fluids. STLE Trans 2004;47(2):182−7.
[9] Born M, Hipeaux JC, Marchand P, Parc G. The relationship between chemical structure and effectiveness of some metallic dialkyl and diaryl dithiophosphates in different lubricated mechanisms. Lub Sci 1992;4:93.
[10] Sethuramiah A, Lubricant performance issues—laboratory Vs real environment, workshop on recent advances in eco-friendly lubricants, IISc, Bangalore, India; 2010.

CHAPTER 3

Dry Wear Mechanisms and Modeling

3.1 INTRODUCTION

Wear may be defined as the progressive loss of material. The wear process results in dimensional changes of the components. There is a tolerance limit to dimensional changes beyond which the performance is drastically affected, necessitating replacement of components. This involves replacement cost as well as loss due to downtime. It has to be kept in mind that in operations like grinding maximizing material removal is what is desired. This involves intentional removal of material and is in a different category.

Wear is being intensively studied over the last six decades. At a fundamental level wear modeling based on mechanisms is being attempted. If such models are effective they can be used for wear prediction and control. Wear is a very complex process and it can be said that modeling of wear based on fundamental principles is not successful. The studies conducted so far however have provided clarity regarding mechanisms that can be useful in material selection. The control of wear with surface coatings is industrially successful and utilizes the knowledge base available at the fundamental level.

This chapter deals with metallic materials. The Hertzian stresses and contact temperatures that are of importance both for dry and lubricated wear are considered first. This is followed by a consideration of adhesive, abrasive, and fatigue wear in separate sections. Oxidative wear is not covered here as it is treated in the chapter on chemical wear. Surface coatings for wear reduction are not covered as they are not within the scope of this book. Wear of nonmetals has not been covered separately as the main interest in chemical wear is with metallic materials. However where appropriate the wear behavior of ceramics has been considered. There are less common wear mechanisms like fretting that are not considered here.

Modeling of Chemical Wear.
DOI: http://dx.doi.org/10.1016/B978-0-12-804533-6.00003-2
© 2016 Elsevier Inc.
All rights reserved.

3.2 HERTZIAN CONTACT STRESSES

The contact of counter formal surfaces is important in engineering contacts like gears and ball bearings. These are also called concentrated contacts. When two spheres are brought into contact the initial point contact expands with load and the stresses involved are high. Asperities also are treated as spherical with a given radius for analyzing stresses. Similarly in line contacts like between discs and gears the contact starts with a line and expands with load into a contact patch. The stress distribution and the governing equations are given below for point and line contacts. These governing equations are for ideally smooth surfaces. When the contact is between rough surfaces the contact dimensions increase and the maximum pressure decreases in comparison to smooth surfaces. However when roughness is low the effects are not significant and smooth surface theory is assumed. This is the case for many concentrated contacts in rolling contact bearings. The estimation of real area with rough surfaces is a complex problem and is not considered in this book.

3.2.1 Stresses in Point Contact

As we can see in Figure 3.1, in (a), a circular patch of radius "a" develops, and the pressure distribution is semielliptical as shown in expanded scale in (b). Maximum stress occurs in the center. In many heavily loaded

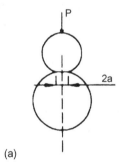

(a)

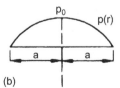

(b)

Figure 3.1 Stress distribution in point contact. (a) Spheres in elastic contact and (b) semielliptical pressure distribution.

contacts the maximum stress can exceed the yield stress of the material. The equations that govern pressure, subsurface shear stress, and tensile stress at the contact periphery are given as

$$\text{Contact radius, } a = \left[\frac{3PR}{4E}\right]^{1/3} \tag{3.1}$$

where P is load, E is reduced modulus, and R is reduced radius. The elliptical pressure distribution is

$$p(r) = p_0\left[1 - \frac{r^2}{a^2}\right]^{1/2} \tag{3.2}$$

where r is any radius within the contact circle. The maximum pressure[1] p_0 occurs on the axis of symmetry and is

$$p_0 = \frac{3}{2}\frac{P}{\pi a^2}$$

The mean pressure $p_m = \frac{2}{3}p_0$

Another important stress is the tensile stress that acts on the surface at the contact periphery and has a value of $k'p_0$ with

$$k' = \frac{\pi f(4 + \nu)}{8} + \frac{(1 - 2\nu)}{3} \tag{3.3}$$

where f is friction coefficient and ν is the Poisson ratio.

The reduced modulus and reduced radius are obtained as in Eqns (1.11) and (1.27) given in Chapter 1. The maximum shear stress for frictionless condition is $0.31\ p_0$ and is located at a depth of $0.48a$ for the materials having a Poisson ratio of 0.3 on the axis of symmetry. When there is friction subsurface stresses get modified with maximum stress point moving upward.

Example 3.2.1.1 Point Contact Stresses

Consider a case of a steel ball of 5 mm radius loaded against a flat of the same material under a load of 200 N. For this material $E_{1,2} = 203$ GPa and $\nu_{1,2} = 0.31$.

---

[1] In Section 3.2.1 of LWST (Lubricated wear — science and technology, Elsevier, 2003) the maximum pressure equation at page 69 has been interchanged by maximum pressure equation at page 70 by mistake.

We find reduced modulus $E = 112.3$ GPa and reduced radius $R = 5 \times 10^{-3}$ m. So the radius of contact $a = 1.88 \times 10^{-4}$ m $= 0.188$ mm.

$$\text{Maximum pressure } p_0 = 2.69 \text{ GPa.}$$

The maximum pressure easily exceeds the yield pressure of medium carbon steel. Also the maximum subsurface shear stress will be equal to $0.31 \times 2.69 = 0.835$ GPa and is of the order of the yield strength. This calculation shows that very high stresses are involved in concentrated contacts. Tensile stresses are involved at the surface even when there is no traction but are significant when sliding is involved. When the ball moves along the surface with a typical friction coefficient f of 0.4 the tensile stress at the contact periphery will be $k'p_0$ and can be calculated from Eqn (3.3). This value will be 2.162 GPa. Such high tensile stresses can induce surface cracks and fatigue wear. These tensile stresses act at the trailing edge of the contact. The leading edge is subject to compressive stresses.

3.2.2 Stresses in Line Contact

The line contacts also involve high stresses with semielliptical distribution of stress. The stress distribution on the surface is given in Figure 3.2, followed by the stress equations.

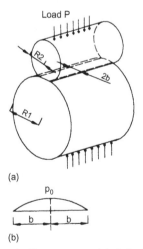

(a)

(b)

Figure 3.2 Stress distribution in line contact. (a) Cylinders in parallel contact and (b) semielliptical pressure distribution.

For two parallel cylinders in static contact P is defined as load per unit length and b is half contact width and is expressed as

$$\text{Half contact width}^2 \; b = \left[\frac{4PR}{\pi E}\right]^{1/2}.$$

$$\text{Maximum pressure } p_0 = \frac{2P}{\pi b} \text{ and mean pressure } p_m = \frac{\pi}{4}p_0.$$

The value of maximum shear stress is $0.30p_0$ and occurs at a depth of $0.78b$ for the materials having a Poisson ratio of 0.3 on the axis of symmetry. The maximum tensile stress acts at the trailing edge with a value of $2fp_0$ where f is coefficient of sliding friction. Since maximum pressures are high the tensile stresses can be substantial. Note tensile stresses act only in a sliding friction situation in line contact.

Example 3.2.2.1 Line Contact Stresses

Consider a spur pinion with 14 teeth transmitting power P_w of 15 kW at 2500 rpm to spur gear with 49 teeth. If the pressure angle φ is 25° and module m is 4 mm, the size of pinion and gear, linear velocity, and normal load may be calculated as follows:

The module m is defined as the ratio of pitch diameter to number of teeth. Therefore the diameter of pinion $D_p = m \times$ number of teeth in pinion $= 4 \times 14 = 56$ mm. Hence the radius of pinion $r_p = 28$ mm. Similarly radius of gear $r_g = 4 \times 49/2 = 98$ mm. Velocity $v = \pi \times D_p \times$ rpm/60 $= 7.33$ m/s. The tangential load $F_t = P_w/v = 15{,}000/7.33 = 2046.4$ N. The normal load W acting on teeth $= F_t/\cos \varphi = 2046.4/\cos 25 = 2257.9$ N. This normal load acts along the pressure line, which is fixed if the teeth are of involute shape and gears have no vibration disturbing the position of centers of gears.

Figure 3.3 shows the pressure line, pitch circle, base circle, and addendum circle of pinion and gear. The tooth engagement starts at A_1 and ends at B_1. At start and end of engagement there is maximum relative sliding while at pitch point O it is pure rolling. So the gear contact is rolling/sliding contact.

At pitch point the radius of curvature of gear tooth is BO taking B as center, and radius of curvature of pinion tooth is AO taking A as center.

---

[2] In Section 3.2.1 of LWST the half contact width equation has been wrongly written. The correct equation is given here.

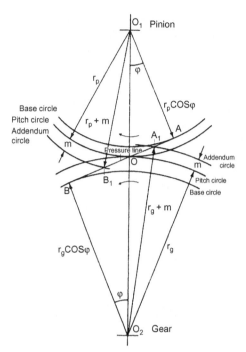

Figure 3.3 The pressure line, base, pitch, and addendum circles of pinion and gears having centers O_1 and O_2, respectively.

At pitch point we may assume two cylinders having radius BO and AO with length equal to face width are in line contact. The BO may be calculated as $r_g \sin \varphi = 98 \sin 25 = 41.42$ mm, and $AO = r_p \sin \varphi = 28 \sin 25 = 11.83$ mm.

Similarly at the start of engagement A_1 we may assume that two cylinders having radius BA_1 and AA_1 with length equal to face width are in contact. The OA_1 may be calculated as $\sqrt{(r_g+m)^2 - (r_g \cos \varphi)^2} - r_g \sin \varphi = 8.73$ mm. Therefore $BA_1 = BO + OA_1 = 41.42 + 8.73 = 50.15$ mm, and $AA_1 = AO - OA_1 = 11.83 - 8.73 = 3.1$ mm.

If both pinion and gear are made of steel having $E_{1,2} = 206$ GPa and $\nu_{1,2} = 0.30$, we find reduced modulus $E = 113.19$ GPa.

The reduced radius R at pitch point $= 9.2$ mm while at start of engagement $= 2.9$ mm. If the face width $= 12.5 \times m$ then its value is $12.5 \times 4 = 50$ mm. The normal load per unit length $P = W/(\text{face width}) = 2257.9/(50 \times 10^{-3}) = 45,158$ N/m.

Table 3.1 Hertzian parameters at pitch point and start of engagement of teeth

Hertzian parameters	At pitch point (zero relative sliding velocity)	At start of engagement (maximum relative sliding velocity)
Half contact width, b	0.068 mm	0.038 mm
Maximum pressure, p_0	0.423 GPa	0.757 GPa
The maximum subsurface shear stress	0.127 GPa	0.227 GPa
The location of maximum shear stress below the surface	0.053 mm	0.030 mm
The maximum tensile stress at trailing edge if coefficient of friction $f = 0.4$		0.606 GPa

Putting the required values in Hertz equations of line contact the parameters at pitch point and start of engagement are tabulated in Table 3.1.

The maximum tensile stress value at trailing edge is substantial and it can initiate surface cracks that can propagate, leading to fatigue wear and pitting. The f value used is high in this example. In lubricated contacts f is expected to be ~ 0.1.

3.3 CONTACT TEMPERATURE

The heat generated due to friction in tribological contacts causes temperature rise at the surface. These temperatures can affect surface oxidation, lubricant failure, chemical reactions, and a host of related problems. This section considers the theoretical approaches to calculate temperature rise. The first step is to calculate temperature rise considering the surfaces are ideally smooth. The next step is to consider the more realistic situation of asperity contacts through which the heat flows into the surfaces.

3.3.1 Theoretical Approach for Smooth Surfaces

The main approach has been due to Jaeger [1], who considered in detail the temperature rise for moving and stationary sources. As applied to a tribological situation consider a square contact as illustrated in Figure 3.4.

Body II is moving over body I with a square contact of $4l^2$ and velocity v. With reference to the heat source of half contact dimension l the

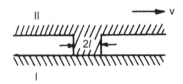

Figure 3.4 Square projection of body II moving over stationary surface I with velocity v.

stationary body I receives heat from a *moving* source while body II receives heat from a *stationary* source.

The nomenclature adopted for the calculation is given here:

k = thermal conductivity (W/m °C)

l = half contact dimension of heat source (m)

q = heat flux (W/m^2)

v = sliding velocity (m/s)

$\Delta\theta$ = average temperature rise (°C)

χ = thermal diffusivity (m^2/s)

$P_e = \frac{vl}{2\chi}$, Peclet number, nondimensional

W = load (N)

The following two equations form the basis for calculating the average temperature rise in contact and are based on simplification of the original theory. For a stationary heat source average temperature rise is given by

$$\Delta\theta = 0.946 \left(\frac{ql}{k}\right) \tag{3.4}$$

This equation is also applicable for slow speed contact with $P_e < 0.1$. Some authors have considered the equation is applicable up to $P_e = 0.5$. For a moving heat source with $P_e > 5.0$ the average temperature rise is

$$\Delta\theta = 0.75 \left(\frac{ql}{k}\right) P_e^{-1/2} \tag{3.5}$$

The case for the Peclet number ranging from 0.1 to 5.0 shall be treated later.

The distribution of temperature in the square contact is not uniform and varies from $-l$ to $+l$. The position of the maximum temperature varies with P_e. The simplified equations refer to an integrated average value over the contact. For stationary contact, the maximum temperature is ≈ 1.15 times the average value while for moving source with $P_e > 5.0$ the maximum value is nearly 1.5 times the average value. The basic equations above refer to heat flow into one body or the other. In sliding contact, the heat generated is distributed into both the bodies II and I. The common interface is considered to have the same temperature. This requirement means that the heat flux shall be so partitioned that $\Delta\theta$ is the same for both bodies. The partition of the heat flux q generated can be handled by the following equation:

$$\frac{1}{\Delta\theta} = \frac{1}{\Delta\theta_I} + \frac{1}{\Delta\theta_{II}} \qquad (3.6)$$

where $\Delta\theta_I$ and $\Delta\theta_{II}$ are temperature rise for body I and body II by considering that *total* heat flux q enters one body or the other.

The case for Peclet numbers in the range of 0.1−5.0 is treated by a graphical procedure developed by Jaeger [1]. The temperature rise for the stationary source as applicable to body II is used as before to obtain $\Delta\theta_{II}$ assuming all heat is flowing to body II. For body I receiving heat from a moving source $\Delta\theta_I$ is calculated from Eqn (3.4) multiplied by a coefficient ε. ε is obtained from

$$\varepsilon = 0.3365yP_e^{-1} \qquad (3.7)$$

where y is a function of Peclet number and is obtained graphically from Figure 3.5.

Tian and Kennedy [2] proposed useful equations for circular heat sources in plastic and elastic contact. The distribution of heat is

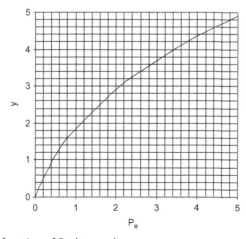

Figure 3.5 "y" as function of Peclet number.

considered uniform for plastic contact. For elastic contact in view of the elliptical pressure distribution the heat distribution is considered nonuniform. The advantage of these equations is that they are generalized for any Peclet number and have a sound theoretical basis. The general approximate equation for average temperature rise at one surface $\Delta\theta$ is

$$\Delta\theta_{1,2} = 2qa\frac{0.61}{k_{1,2}\pi^{0.5}\left(0.6575 + P_{e1,2}\right)^{0.5}} \quad \text{(for plastic contact)} \quad (3.8)$$

$$\Delta\theta_{1,2} = 2qa\frac{0.732}{k_{1,2}\pi^{0.5}\left(0.874 + P_{e1,2}\right)^{0.5}} \quad \text{(for elastic contact)} \quad (3.9)$$

where q is heat flux and "a" is contact radius and applicable for any Peclet number P_e. Subscripts 1,2 refer to one surface or the other. These equations predict temperature very close to the value obtained analytically. The maximum difference between two solutions for the entire range of Peclet numbers was less than 5%. For a stationary case, these equations tend to exact values given as follows:

$$\Delta\theta = \frac{8Q}{3\pi^2 ka} \quad \text{(for plastic contact)} \quad (3.10)$$

$$\Delta\theta = \frac{9Q}{32ka} \quad \text{(for elastic contact)} \quad (3.11)$$

where Q is total heat going through the contact circle, k is thermal conductivity, and "a" is contact radius.

Average temperature at the surface may be calculated using Eqn (3.6) after calculating $\Delta\theta_1$ and $\Delta\theta_2$ from the above relevant equation.

Example 3.3.1.1 Smooth Surface Temperature Rise

The previous procedure can be illustrated by the following example. Consider a stationary hollow circular pin[3] made of ceramic having outer diameter d_{po} of 10.3 mm and inner diameter d_{pi} of 5.5 mm. This pin is in sliding contact with a steel disc rotating at 1000 rpm. The load W is 3 kgf = 29.43 N. The coefficient of friction is 0.42. The wear track formed on the disc has an outer diameter d_{wo} of 97.4 mm and inner

---

[3] In Section 1.4.1 of LWST at page 16, last paragraph, the example is for stationary steel pin in sliding contact with moving glass disc.

Table 3.2 Thermo-mechanical properties of alumina pin and steel disc

Thermo-mechanical property	Steel disc	Alumina pin
Thermal conductivity, k	43 W/m °C	33 W/m °C
Thermal diffusivity, χ	1.172×10^{-5} m²/s	9.615×10^{-6} m²/s
Hardness, H	3.26×10^9 N/m²	1.138×10^{10} N/m²

diameter d_{wi} of 90.2 mm. The thermo-mechanical properties are listed in Table 3.2.

The problem is to find the surface temperature rise based on geometric area of contact. The geometric area of contact A_g is $\pi(d_{po}^2 - d_{pi}^2)/4 = 5.956 \times 10^{-5}$ m².

For the square contact $l = \sqrt{\frac{5.956 \times 10^{-5}}{4}} = 3.86 \times 10^{-3}$ m

The tangential velocity at the mean circle of wear track $v = \frac{\pi[(d_{wo} + d_{wi})/2] \text{ rpm}}{60} = \frac{\pi * 9.38 \times 10^{-2} * 1000}{60} = 4.91$ m/s

The total frictional heat $Q = f\, Wv = 0.42 * 29.43 * 4.91 = 60.69$ W. Therefore heat flux $q = Q/A_g = 60.69/5.956 \times 10^{-5} = 1.02 \times 10^6$ W/m².

The pin receiving heat from the stationary source is called body II while the disc receiving heat from the moving source is called body I.

The Peclet number $P_e = \frac{vl}{2\chi} = \frac{4.91 * 3.86 \times 10^{-3}}{2 * 1.172 \times 10^{-5}} = 808.6$. As $P_e > 5$, Eqn (3.5) is applicable. Assuming all heat is flowing to body I:

$$\Delta\theta_I = 0.75\left(\frac{ql}{k_I}\right)P_e^{-1/2} = 0.75 * \left(\frac{1.02 \times 10^6 * 3.86 \times 10^{-3}}{43}\right)808.6^{-1/2}$$

$$= 2.41 \; °C.$$

Assuming all heat is flowing to body II, Eqn (3.4) is applicable:

$$\Delta\theta_{II} = 0.946\left(\frac{ql}{k_{II}}\right) = 0.946 * \left(\frac{1.02 \times 10^6 * 3.86 \times 10^{-3}}{33}\right) = 112.87 \; °C$$

Now temperature rise $\Delta\theta$ is obtained from Eqn (3.6):

$$\frac{1}{\Delta\theta} = \frac{1}{\Delta\theta_I} + \frac{1}{\Delta\theta_{II}} = \frac{1}{2.41} + \frac{1}{112.87}$$

So, $\Delta\theta = 2.36 \; °C$.

3.3.2 Theoretical Approach for Rough Surface

Let us consider contact between two nominally flat surfaces. The actual contact occurs at asperities. Normally this real area of contact A_r is much lower than geometric contact area A_g. The relationship between real area of contact and constriction resistance has been studied by Holm [3] and Greenwood [4]. If geometric contact area is circular with radius l_g, and has n number of asperities with an average asperity radius a, then constriction resistance R can be found by treating n asperities as parallel resistance coupled with the geometric contact area resistance in series. The relationship is

$$R = \rho \left(\frac{1}{2na} + \frac{1}{2l_g} \right) \tag{3.12}$$

where ρ is resistivity, $\Omega \cdot m$.

Considering the heat flow problem as analogous to the electrical conduction problem, average frictional temperature rise for rough surfaces may be approximated as follows:

$$\Delta\theta = \Delta\theta_a + \Delta\theta_g \tag{3.13}$$

where

$\Delta\theta_a$ = temperature rise due to asperity alone considering heat input Q/n is based on uniform distribution over asperities

$\Delta\theta_g$ = temperature rise over the geometric area considering heat input Q

If asperities are dense a constriction alleviation term should be added to consider the heat flow across metallic junctions [5]. This additional term reduces the temperature due to extraction of heat Q/n from asperity over a circle radius of $l_g/\sqrt{n}$ [6]. So a general expression irrespective of asperity density can be given as follows [7]:

$$\Delta\theta = \Delta\theta_a + \Delta\theta_g - \Delta\theta_{ag} \tag{3.14}$$

where $-\Delta\theta_{ag}$ is negative temperature over circular area of radius $l_g/\sqrt{n}$ considering heat input Q/n.

This equation can be reduced to Eqns (3.15) and (3.16) using Eqns (3.8) and (3.9) for plastic and elastic contacts. These equations are applicable to any Peclet number as stated earlier. With Peclet number zero they are applicable to stationary contacts.

$$\Delta\theta = \frac{0.2191Q}{k}\left(\frac{1}{n\psi_1(a)} + \frac{1}{\psi_1(l_g)} - \frac{1}{n\psi_1(l_g/\sqrt{n})}\right) \quad \text{(for plastic contact)}$$

(3.15)

$$\Delta\theta = \frac{0.2629Q}{k}\left(\frac{1}{n\psi_2(a)} + \frac{1}{\psi_2(l_g)} - \frac{1}{n\psi_2(l_g/\sqrt{n})}\right) \quad \text{(for elastic contact)}$$

(3.16)

where $\psi_1(x) = x\sqrt{0.6575 + \frac{vx}{2\chi}}$ and $\psi_2(x) = x\sqrt{0.874 + \frac{vx}{2\chi}}$

In limiting case, if $A_r = A_g$ then $a = l_g/\sqrt{n}$. Hence the first and third terms in the previous expressions vanish. The temperature rise is only due to geometric area. If $A_r \ll A_g$ then $a \ll l_g/\sqrt{n}$. Therefore the third term may be neglected in Eqns (3.15) and (3.16). In this case temperature rise may be found just from Eqn (3.13).

Thus Eqns (3.15) and (3.16) are for average steady state asperity temperature rise at any asperity density and Peclet number under plastic and elastic contact, respectively. An asperity temperature rise at rough surface may be calculated using Eqn (3.6) after calculating $\Delta\theta_1$ and $\Delta\theta_2$ from the earlier relevant equations depending on the type of contacts. Note that $\Delta\theta_1$ and $\Delta\theta_2$ are temperature rise for body I and body II by considering that *total* heat flux enters one body or the other.

Recently a theoretical multiscale analysis of asperity temperature rise problem was done by Jang and Barber [7]. They considered the N-scale fractal surface defined by Archard [8] as well as cases with more realistic fractal dimensions. They found that estimate of temperature rise in fractal surface increases with increase of scale N but it becomes asymptotic at a particular value. This bound temperature increases with fractal dimension D. The detailed analysis is available in [7]. These ideas at the present stage are difficult to implement and the above analysis based on asperity areas only is retained.

Example 3.3.2.1 Asperity Temperature Rise
Consider Example 3.3.1.1. Actual contact occurs at asperities. First of all the asperity temperature rise without thermal interaction will be

illustrated and subsequently the effect of thermal interaction will be demonstrated.

The asperity temperature rise calculation based on real area of contact in the present example is due to asperities alone. Hence in using Eqn (3.14) the second term $\Delta\theta_g$ and third term $\Delta\theta_{ag}$ are neglected. Calculation is based on plastic flow at asperities. The parameters related to the asperities carry the subscript a in the following text.

The real area of contact is calculated from Eqn (1.1) $A_r = W/H$ where H is hardness of softer material. Therefore $A_r = 29.43/3.26 \times 10^9 = 9.028 \times 10^{-9}$ m^2. The calculated heat flux on the basis of real area $q_a = Q/A_r = 60.69/9.028 \times 10^{-9} = 6.72 \times 10^9$ W/m^2.

Assume an average half contact dimension l_a for the square asperity of 5 μm. The number of asperities n comes around to be 90. If the asperity contact is assumed to be circular and its radius to be 5 μm, n comes around to be 115. In view of the above, n is assumed to be 100 and recalculations of asperity size was done as follows. The asperity contact area A_a is 9.028×10^{-11} m^2. The asperity temperature now may be calculated either using Jaeger's approach [1], where asperity is considered square, or Tian and Kennedy's approach [2], considering circular asperity. The parameters calculated for the above are tabulated in Table 3.3.

The results in Table 3.3 show that asperity temperature rise of 294.7 °C estimated by using Tian and Kennedy's equation considering circular asperity is very close to 296.8 °C estimated by using Jaeger's approach considering square asperity.

For elastic contact if mean pressure is 0.4 H then a and q_a will be 8.48×10^{-6} m and 2.69×10^9 W/m^2, respectively. Using Eqn (3.9) for elastic contact, the asperity temperature rise calculated is 186.8 °C. It is substantially lower than 294.7 °C for the plastic contact. Well-run engineering surfaces have mainly elastic contacts and this is advantageous from a temperature point of view. Contact models described in Chapter 1 will be useful to estimate mean pressure and asperity size for elastic contacts.

Now Tian and Kennedy's Eqn (3.8) will be used to demonstrate the effect of thermal interaction on asperity temperature rise as the equation is applicable for any Peclet number.

Consider the same example discussed earlier in this section and the previous section. The real area of contact A_r is 9.028×10^{-9} m^2, assuming number of asperities n = 100. The contact radius of circular asperity $a = \sqrt{A_a/\pi} = 5.36 \times 10^{-6}$ m.

Table 3.3 Asperity temperature rise without thermal interaction

Parameters	Jaeger's approach [1]	Tian and Kennedy's approach [2]
Asperity contact dimension	Half contact dimension of square asperity $$l_a = \sqrt{\frac{A_a}{4}} = \sqrt{\frac{9.028 \times 10^{-11}}{4}}$$ $$= 4.75 \times 10^{-6} \text{ m}$$	Contact radius of circular asperity $$a = \sqrt{\frac{A_a}{\pi}} = \sqrt{\frac{9.028 \times 10^{-11}}{\pi}}$$ $$= 5.36 \times 10^{-6} \text{ m}$$
Peclet number	$$P_{ea} = \frac{v l_a}{2\chi} = 0.995$$	$$P_{ea} = \frac{v a}{2\chi} = 1.123$$
$\Delta\theta_I$ (assuming all heat is flowing in body I; i.e., steel disc and applying moving heat source theory as heat source is moving with respect to body I)	$\Delta\theta_I = 0.946 \left(\frac{q_a l_a}{k_I}\right)\varepsilon$ as $0.1 < P_{ea} < 5$. ε is found from Eqn (3.7) $\varepsilon = 0.3365 y P_{ea}^{-1}$. y is obtained from Figure 3.5, which is 1.85. Therefore $\Delta\theta_I = 439.4\,^\circ$C	Since plastic flow is considered at asperities hence Eqn (3.8) for uniform heat distribution is used $$\Delta\theta_I = 2q_a \frac{0.61}{k_I \pi^{0.5}(0.6575 + P_{ea})^{0.5}}$$ Note that Peclet number P_{ea} in this case is 1.123 and Eqn (3.14) becomes Eqn (3.8) on neglecting $\Delta\theta_g$ and $\Delta\theta_{ag}$ Therefore $\Delta\theta_I = 432.1\,^\circ$C
$\Delta\theta_{II}$ (assuming all heat is flowing in body II; i.e., ceramic pin and applying stationary heat source theory as heat source is stationary with respect to body II)	$$\Delta\theta_{II} = 0.946 \left(\frac{q_a l_a}{k_{II}}\right)$$ $$= 915.0\,^\circ\text{C}$$	$$\Delta\theta_{II} = 2q_a \frac{0.61}{k_{II} \pi^{0.5}(0.6575 + P_{ea})^{0.5}}$$ $$= 926.5\,^\circ\text{C}$$ Note that Peclet number P_{ea} in this case is 0 and Eqn (3.14) becomes Eqn (3.8) on neglecting $\Delta\theta_g$ and $\Delta\theta_{ag}$
Asperity temperature rise, $\Delta\theta$	using Eqn (3.6) $$\Delta\theta = (\Delta\theta_I^{-1} + \Delta\theta_{II}^{-1})^{-1}$$ $$= 296.8\,^\circ\text{C}$$	using Eqn (3.6) $$\Delta\theta = (\Delta\theta_I^{-1} + \Delta\theta_{II}^{-1})^{-1}$$ $$= 294.7\,^\circ\text{C}$$

The geometric contact area A_g is $5.956 \times 10^{-5}\,m^2$. Considering geometric contact as solid circular, the contact radius $l_g = \sqrt{A_g/\pi} = 4.35 \times 10^{-3}\,m$.

As plastic flow is considered at asperities, Eqn (3.15) is used for asperity temperature rise calculations.

Consider body I, the steel disc. The heat source moves with velocity $v = 4.91\,m/s$ with respect to body I. Assuming all heat is flowing into body I:

$$\Delta\theta_I = \frac{0.2191Q}{k_I}\left(\frac{1}{n\psi_1(a)} + \frac{1}{\psi_1(l_g)} - \frac{1}{n\psi_1(l_g/\sqrt{n})}\right)$$

The values of function $\psi_1(x) = x\sqrt{0.6575 + vx/2\chi}$ in this equation is calculated as follows:

$$\psi_1(a) = a\sqrt{0.6575 + va/2\chi} = 7.15 \times 10^{-6}\,m$$

$$\psi_1(l_g) = l_g\sqrt{0.6575 + vl_g/2\chi} = 1.32 \times 10^{-1}\,m$$

$$\psi_1(l_g/\sqrt{n}) = (l_g/\sqrt{n})\sqrt{0.6575 + v(l_g/\sqrt{n})/2\chi} = 4.17 \times 10^{-3}\,m$$

therefore

$$\Delta\theta_I = \frac{0.2191 * 60.69}{43}\left(\frac{1}{100 * 7.15 \times 10^{-6}} + \frac{1}{1.32 \times 10^{-1}}\right.$$
$$\left. - \frac{1}{100 * 4.17 \times 10^{-3}}\right)$$
$$= 434.1\,°C$$

If the constriction alleviation term $\Delta\theta_{ag}$ (i.e., third term in the above equation) is neglected then $\Delta\theta_I$ is $434.8\,°C$.

Consider body II, the ceramic pin. The heat source is stationary with respect to body II. Assuming all heat is flowing in body II:

$$\Delta\theta_{II} = \frac{0.2191Q}{k_{II}}\left(\frac{1}{n\psi_1(a)} + \frac{1}{\psi_1(l_g)} - \frac{1}{n\psi_1(l_g/\sqrt{n})}\right)$$

The values of function $\psi_1(x) = x\sqrt{0.6575 + vx/2\chi}$ in the above equation is calculated by putting $v = 0$ as the heat source is stationary for this case. The calculations are

$$\psi_1(a) = a\sqrt{0.6575} = 4.35 \times 10^{-6} \, \text{m}$$

$$\psi_1(l_g) = l_g\sqrt{0.6575} = 3.53 \times 10^{-3} \, \text{m}$$

$$\psi_1(l_g/\sqrt{n}) = (l_g/\sqrt{n})\sqrt{0.6575} = 3.53 \times 10^{-4} \, \text{m}$$

therefore

$$\Delta\theta_{II} = \frac{0.2191 * 60.69}{33} \left(\frac{1}{100 * 4.35 \times 10^{-6}} + \frac{1}{3.53 \times 10^{-3}} \right.$$
$$\left. - \frac{1}{100 * 3.53 \times 10^{-4}} \right)$$
$$= 1029.0 \, °C$$

If the constriction alleviation term $\Delta\theta_{ag}$ (i.e., third term in the above equation) is neglected then $\Delta\theta_{II}$ is 1040.5 °C.

Now asperity temperature rise $\Delta\theta$ is obtained from the Eqn (3.6).

$$\frac{1}{\Delta\theta} = \frac{1}{\Delta\theta_I} + \frac{1}{\Delta\theta_{II}} = \frac{1}{434.1} + \frac{1}{1029.0}$$

So, $\Delta\theta = 305.3 \, °C$.

If the constriction alleviation term $\Delta\theta_{ag}$ (i.e., third term in Eqn (3.14)) is neglected then $\Delta\theta$ is 306.7 °C. The difference in estimate is only 0.46% since real area is a very small fraction of geometric area. The constriction alleviation term will be significant in asperity temperature rise estimation if asperities are dense.

It should be noted that contact temperature T_c responsible for surface oxidation, chemical reactions, lubricant failure, and related problems is obtained by adding the bulk temperature T_b to the estimated asperity temperature rise $\Delta\theta$. The contact temperature T_c is expressed as

$$T_c = T_b + \Delta\theta \qquad (3.17)$$

The bulk temperature T_b is assumed to be same for both contacting bodies. This assumption is valid in lubricated contact in which oil bath temperature may be taken as bulk temperature [9]. The main reason for this is the convective cooling due to the oil that tends to equalize the temperatures. The dry contact situation is more complex and the bulk temperatures can be different for both surfaces, and the problem needs a complete heat transfer analysis using proper boundary conditions [10,11]. The other complications that arise include variations in thermal properties due to temperature as well as influence of oxide and other films on heat transfer [12].

It needs to be emphasized that the above treatments for temperature rise are based on steady state or quasisteady state. Many investigators assumed steady state conditions would prevail. The reasons for this assumption as elucidated by Kennedy [13] are

1. The steady state assumption leads to estimation of maximum possible temperature and hence leads to a conservative estimate.
2. The steady state is usually reached in very short times and in many cases the assumption is justified.

As discussed in the earlier reference and based on the early work of Jaeger and later simplifications, a sliding distance of the order of 1.25 times the length of the heat source is adequate to reach the steady state. This justifies the second reason. However there may be cases where transient analysis may be necessary as in high speed finishing and grinding operations where contact times are very short.

Remember that temperature rise estimates in present analysis is average temperature rise. The temperature distribution within the contact is not uniform. It is difficult to decide whether average rise or maximum rise governs the tribological behavior.

3.4 ADHESIVE WEAR

3.4.1 Phenomena

Adhesion and transfer of material can occur at asperity contacts. The transferred material is consequently removed from surfaces after a number of cycles. Such a wear process is called adhesive wear. The concept of adhesive wear has its origins in the plastic contact model of Bowden and Tabor as discussed in Section 1.2.2.1 in Chapter 1. The junctions formed at asperities break in the sliding process leading to transfer of material. The break of junction may occur in the softer metal or at the interface depending on the shear strength of the interface. Larger transfer occurs when the interface is stronger than the softer material. Occasionally the harder material also can get transferred. The particles that are transferred grow with repeated contacts and eventually break into loose debris after fracture. During the wear process there can be back transfer of material also in some cases. These steps involved in the wear process have been well demonstrated in the earlier work [14]. The junctions involve intermetallic bonding and can have varying strength. While adhesion involving van der Waals forces occurs between any two surfaces the word adhesive in the present context involves metallic bonding and particle transfer. The transfer involved is particulate and the mechanisms involved are normally treated at this level. Some present efforts are involved to understand wear process at the atomic level in micro and nano contacts [15].

Two types of adhesive wear may be distinguished. The first one involves adhesion with no oxide film on the surface. Such situations are involved in ultrahigh vacuum. As discussed in Chapter 1 such contacts can have significant growth in contact area and seize rather than wear. The second situation involves oxide-covered surfaces through which some adhesion occurs. This adhesion leads to transfer and wear. The transfer particles can get easily oxidized and may be removed as oxide particles. The wear may attain a steady state and was related to a balance between new surface generation and oxidation. This balance can get upset when operating severity increases, leading to large-scale adhesion and possible seizure. This damage, termed scuffing, is unacceptable in engineering situations. Effective experimental justification of these ideas was provided in the 1960s by Welsh [16]. This work clearly demonstrated the protective role of oxides in wear. The problem is of particular importance in lubricated contacts and is dealt with in a later chapter. It is also possible that oxides may in some cases totally prevent adhesive transfer. In such cases the wear will be governed by the oxide removal and formation. The usual process of wear with oxide-covered surfaces is illustrated in Figure 3.6.

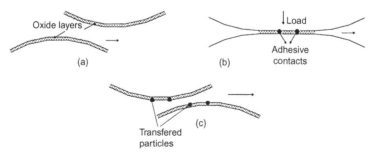

Figure 3.6 Illustration of adhesive transfer through oxide layers at asperity level: (a) approach, (b) contact, and (c) disengagement.

The problem of adhesion in metal forming operations like cold rolling is of a special nature. Cold rolling is a typical example of such a process where reduction in each pass generates large areas of fresh metal surfaces. Despite lubrication adhesive transfer can occur on to the rolls that in turn affects the quality of the rolled product. Aluminum rolling is an interesting example as the surfaces involved are prone to easy break down of oxides and significant adhesion. Methods to improve the situation include coatings and special additives. The solutions so far are based on empirical methods. The knowledge of adhesion at the atomic level is growing at a rapid pace [17] and interesting attempts are now being made to link this knowledge to practice [18].

3.4.2 Modeling of Adhesive Wear

The previous subsection described the complexity involved in adhesive wear. Despite this a wear model was proposed by Archard [19] in the early 1950s. The model and the resulting equation are first described. A critical appraisal of the model then follows.

The model starts with the assumption that the real contact area is plastic and as described in Eqn (1.1) the real area $A_r = \frac{W}{H}$ where W is the load and H is the hardness.

Assume an average asperity contact diameter of d. If it is further assumed that the wear particle generated at the asperity over a distance d is hemispherical its volume will be

$$v_d = \frac{1}{12}\pi d^3 \tag{3.18}$$

total asperities in contact n is given by

$$n = \text{real area/individual asperity area} = \frac{W}{H}\frac{4}{\pi d^2} \tag{3.19}$$

Total wear volume V_d over distance d is then

$$V_d = \frac{1}{3}\frac{Wd}{H} \tag{3.20}$$

If wear rate V_r is now defined as wear volume per unit sliding distance

$$V_r = \frac{V_d}{d} = \frac{1}{3}\frac{W}{H} \tag{3.21}$$

The wear volume V is normally measured and known over a sliding distance l. In these terms

$$V_r = \frac{\sum V_d}{\sum d} = \frac{V}{l} = \frac{1}{3}\frac{W}{H}$$

(3.22)

Many experimental observations showed that the actual wear rates are far lower than obtained by the above equation. To obtain realistic values Archard introduced a nondimensional wear coefficient K and the final equation is

$$V_r = K\frac{W}{H}$$

(3.23)

The constant 1/3 in Eqn (3.22) now gets included in K. The value of K can vary over a wide range of 10^{-2} to 10^{-6}. Archard proposed that K may be considered as the probability that a given encounter results in a wear particle. The value of K can only be obtained experimentally.

There is no satisfactory explanation for K. It is difficult to see why only one in several thousand contacts results in a wear particle. One possible reason for small values of K is the influence of oxide film. If it is assumed that adhesion occurs through defects in oxide film and is rate determining, a plausible explanation can be given as described below.

Assume that in an asperity contact a small circular contact only makes adhesive contact through the oxide film. Further consider that *every* metal contact results in adhesive wear. If α is the fraction of real area that is metallic then diameter of the adhesive particle involved in the transfer will be $d\alpha^{0.5}$. Consider *all* events result in wear particles. Each wear particle is generated over sliding distance d as in Archard's model. It can now be shown that the wear rate is

$$\frac{V}{l} = \alpha^{1.5}\left(\frac{W}{H}\right)\left(\frac{1}{3}\right)$$

(3.24)

As per the above equation, K, the wear coefficient, will be equal to $\alpha^{1.5}/3$. Thus if $\alpha = 0.01$ the wear coefficient will be 3.3×10^{-4}. But now the coefficient has a better physical basis as *every* adhesive junction results in a wear particle. However K cannot be obtained from fundamental considerations and can only be obtained experimentally.

The major contribution of Archard is the concept of wear coefficient. This coefficient is very useful in comparing the wear rates of materials, whatever may be the detailed wear mechanisms. It is also interesting to know that experimentally many wear situations over a range of operating conditions can be fitted to the wear equation with constant K. This will not apply when transitions occur.

Detailed mechanisms have been studied by various authors and modified models have been proposed. For example, Kato [20] proposed that adhesive wear occurs only when deformation is plastic and models the wear process. Finkin [21] also considered that only plastic contacts contribute to adhesive wear and estimated wear coefficients. Several additional complications can occur due to microstructural modification at the surface [22], debris compaction in the contact zone [23], and work hardening [24]. The way the surface oxides behave is also an important consideration in adhesive wear.

These complexities prompted some researchers to define adhesive wear simply as sliding wear. It is divided into mild and severe ranges. Mild wear involves wear coefficients that are less than 10^{-4} while severe wear involves coefficients greater than 10^{-3}.

3.5 ABRASIVE WEAR

3.5.1 Phenomena of Abrasive Wear

Abrasive wear, as the name suggests, is caused by the cutting action of abrasive material. A good example of the process is grinding of material in which grinding particles remove material. Abrasion can occur also when hard asperities plough through softer material. This type of abrasive wear is termed two-body since one of the sliding bodies generates abrasive wear. Sometimes abrasion is caused by extraneous hard particles coming between the sliding pair of materials. This can occur when dust particles get into the lubrication circuit. Trapped wear particles that oxidize and harden can also act as abrasive third bodies.

Early research was focused on two-body abrasive wear [25–27]. The major findings of this work may be summarized as follows:

- Abrasive wear resistance of pure metals is nearly proportional to hardness.
- In the case of work hardening materials the wear resistance corresponds to that of the work hardened material.
- For alloys, including steel, wear resistance is less than proportional to hardness and depends on the microstructure.
- Abrasive wear can occur when the hardness of the abrasive is 1.0–1.6 times that of the material. No abrasive wear is expected below this range.

- All the grooved material need not be removed by cutting action. Part of the deformed material can form a wedge on the surface.
- The three-body abrasive wear is of industrial importance but less investigated. This is partly because of the ill-defined nature of this wear. The size, hardness, and particle morphology are important factors. Also the particle size in relation to oil film thickness is important as this decides whether a particle is drawn into the contact.

3.5.2 Modeling of Abrasive Wear

Two-body abrasive wear has been modeled on the basis of volume of groove formed when a conical indenter is slid. It is assumed ideally that there is only cutting and all the grooved material is removed. But in reality the abrasive grain may have morphology far different from a cone. Also it is known that the abrasive action may involve both plastic deformation and cutting. The types of wear are illustrated in Figure 3.7 and based on the paper of Kato [28].

From Figure 3.7 it can be seen there can be ploughing alone with no wear, cutting action coupled with wedge formation, and dominant cutting with a low level of wedge formation. All these factors affect wear rate and following the argument of Archard [29] a probability factor may be introduced. Without going into details the final form of the equation may be expressed as

$$\frac{V}{l} = K_{ab}\frac{W}{H} \tag{3.25}$$

This equation is similar to that of adhesive wear and once again the wear coefficient cannot be derived from fundamental considerations.

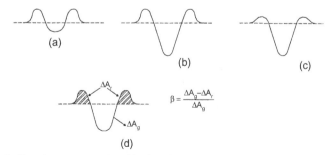

Figure 3.7 Abrasive wear modes: (a) ploughing, (b) wedge action, (c) cutting, and (d) definition of β based on cross-sectional area of ridge and groove. Source: *Based on [28]*.

However the factors that govern the wear factor are known and this helps in selecting and developing abrasive wear resistant materials and coatings. Also better wear modeling is being attempted and these equations tend to be material-specific. An interesting example is the development of a model for composite material [30].

3.6 FATIGUE WEAR

3.6.1 Phenomena and Models

Fatigue and fracture are well-known phenomena in material science. The central ideas are recalled here. When a notched bar with a crack length a_c is subject to tensile stress, a stress intensity factor may be defined as

$$K_I = B\sigma(\pi a_c)^{0.5} \tag{3.26}$$

where B is the geometric factor and σ is the applied stress.

When a critical stress intensity factor K_{Ic} is reached the crack propagates very quickly, leading to fracture. K_{Ic} is called the fracture toughness of the material expressed in $N \cdot m^{-3/2}$.

When the stress intensity is lower than critical and the material is cyclically stressed, at a given stress there will be incremental growth in crack length at each cycle. When the crack length grows to a level where fracture toughness is reached, fracture occurs. The process of incremental growth in crack is referred to as fatigue.

Fatigue wear refers to the material removal at the asperity level or at the Hertzian contact dimension level due to repeated stress cycles. For elastic contacts the process involved is akin to high cycle fatigue. As discussed in Section 3.2, in concentrated contacts high tensile stresses can occur at the contact edge as well as the asperities. If a line contact is considered the maximum tensile stress occurs at the trailing edge of the contact and is equal to $2fp_0$. These stresses are high and can induce surface cracks that propagate, resulting in a wear particle after a number of cycles. Each particle generation may be considered a micro fracture. In dry contacts this mode of wear is common in ceramics because of their low fracture strength. Also in some situations fatigue wear is possible in steel contacts. Karmakar [31] modeled wear in boundary-lubricated rolling/sliding contacts assuming that the tensile stresses are responsible for fatigue wear. This model developed a criterion for fatigue by an independent cyclic stressing method and then applied it to a rolling/sliding contact. More recently Laine et al. have related micropitting in lubricated contacts to tensile stresses [32].

Although fatigue wear can be identified from surfaces and wear particles in some cases, in many others such clarity is not possible as wear may be due to a combination of several mechanisms. In such cases we can discuss or model wear only in terms of the Archard equation and determine the wear coefficient experimentally.

There can be fatigue wear with low cycle fatigue. This mode is related to repetitive plastic deformation at the surface/subsurface and is less common. For example, it can occur in abrasive wear when the plastically deformed material is repeatedly stressed. It can also occur in rail-wheel contacts.

Another form of fatigue widely investigated is that of pitting. The investigations are mainly with lubricated line and point contacts. Pitting is also referred to as fatigue wear but the authors would prefer to call it a failure. Pitting involves formation of deep pits that make the contact nonfunctional. The early work on pitting showed that this damage is related to the subsurface orthogonal shear stress. Several experiments showed that fatigue (pitting) life is related to (load)^n where n is around 3. Bearing selection is based on such criteria. For line contacts like gears the pitting life follows a similar approach with a different exponent. The later research indicates that pitting is due to a combination of subsurface shear stresses and surface asperity interactions [33]. The surface and subsurface cracks and their propagation together result in pitting. Thus if there is asperity contact through EHL film the pitting life will be lower.

NOMENCLATURE

a	contact radius in point contact
a	average asperity contact radius
a_c	crack length
A_a	asperity contact area
A_g	geometric area of contact
A_r	real area of contact
b	half contact width in line contact
B	geometric factor
d	average asperity contact diameter
d_{pi}, d_{po}	inner and outer diameters of hollow circular pin
d_{wi}, d_{wo}	inner and outer diameters of wear track formed on the disc
D	fractal dimension
D_p	pitch diameter of pinion

E	effective elastic modulus of the two bodies in contact
E_1, E_2	elastic moduli of the two materials 1 and 2
f	coefficient of friction
F_t	tangential load in gear transmission
H	hardness of softer material
k	thermal conductivity
k'	multiplication factor to obtain tensile stress at circular contact periphery
k_I, k_{II}	thermal conductivities of bodies I and II
$k_{1,2}$	thermal conductivity of material 1 or 2
K	adhesive wear coefficient, nondimensional
K_{ab}	abrasive wear coefficient, nondimensional
K_I	stress intensity factor
K_{Ic}	critical stress intensity factor
l	half contact dimension of heat source
l	sliding distance
l_a	half contact dimension of square asperity
l_g	geometric contact radius
m	module, ratio of pitch diameter to number of teeth in a gear
n	number of asperities
N	scale of fractal surface
p_0	maximum pressure
p_m	mean pressure
$p(r)$	elliptical pressure distribution
P	load in point contact
P	load per unit length in line contact
P_e	$= \frac{vl}{2\chi}$, Peclet number, nondimensional
P_{ea}	Peclet number for asperity contact
$P_{e1,2}$	Peclet number for body 1 or 2 as applicable
P_w	power transmitted in gear mechanism
q	heat flux
q_a	heat flux on the basis of real area
Q	total heat going through contact circle
r	any radius within the contact circle of radius a
r_p, r_g	pitch radius of pinion and gear
R	equivalent radius of two bodies in contact
R	constriction resistance defined as $R = \rho\{1/(2na) + 1/(2l_g)\}$
R_1, R_2	radii of two bodies in contact
T_b	bulk temperature
T_c	contact temperature
v	pitch velocity of gear
v	tangential velocity at the mean circle of wear track
v	velocity of the moving heat source
v_d	wear volume removed at one contact over a sliding distance d
V_d	wear volume generated at all asperity contacts over a sliding distance d
V_r	wear rate, wear volume/sliding distance
W	normal load
y	ordinate value in Figure 3.5

Greek Letters

α fractional metal contact area within real area

β $= \frac{\Delta A_g - \Delta A_r}{\Delta A_g}$ as in Figure 3.7

χ thermal diffusivity

$\Delta\theta$ average temperature rise in contact

$\Delta\theta_a$ asperity temperature rise in contact

$-\Delta\theta_{ag}$ constriction alleviation term in heat flow when asperities are dense

$\Delta\theta_g$ temperature rise in the geometric contact area

$\Delta\theta_I$ surface temperature rise assuming all heat is flowing to body I due to square heat source

$\Delta\theta_{II}$ surface temperature rise assuming all heat is flowing to body II due to square heat source

$\Delta\theta_1$ surface temperature rise assuming all heat is flowing to body I due to circular heat source

$\Delta\theta_2$ surface temperature rise assuming all heat is flowing to body II due to circular heat source

ε coefficient $= 0.3365 y P_e^{-1}$

φ pressure angle in gear transmission system

ν Poisson ratio

ν_1, ν_2 Poisson ratio of materials 1 and 2

ρ resistivity

σ applied stress

$\psi_1(x)$ a function defined as $x\sqrt{0.6575 + vx/2\chi}$

$\psi_2(x)$ a function defined as $x\sqrt{0.874 + vx/2\chi}$

REFERENCES

[1] Jaeger JC. Moving sources of heat and the temperature at sliding contacts. Proc R Soc NSW 1942;76:203−24.

[2] Tian X, Kennedy FE. Maximum and average flash temperatures in sliding contacts. J Trib ASME Trans 1994;116:167−74.

[3] Holm R. Electric contacts handbook. Berlin: Springer-Verlag; 1958.

[4] Greenwood JA. Constriction resistance and the real area of contact. Brit J Appl Phys 1966;17:1621−32.

[5] Hunter A, Williams A. Heat flow across metallic joints—constriction alleviation factor. Int J Heat Mass Transfer 1969;12:524−6.

[6] Burton RA, Burton RG. Cooperative interactions of asperities in the thermotribology of sliding contacts. IEEE Trans Components Hybrids Manufact Tech 1991;14:23−5.

[7] Jang YH, Barber JR. Multiscale analysis of moving clusters of microcontacts. Int J Heat Mass Transfer 2010;53:3817−22.

[8] Archard JF. Elastic deformation and the laws of friction. Proc R Soc Lond A 1957;243(1233):190−205.

[9] Dinc OS, Ettles CM, Calabrese SJ, Scarton HA. The measurement of surface temperature in dry or lubricated sliding. J Trib ASME 1993;115(1):78−82.

[10] Barber JR. The conduction of heat from sliding solids. Int J Heat Mass Transfer 1970;13:857−69.

[11] Tian X, Kennedy FE. Contact surface temperature models for finite bodies in dry and boundary lubricated sliding. J Trib ASME 1993;115:411−18.

[12] Berry GA, Barber JR. The division of friction heat—a guide to the nature of sliding contact. J Trib ASME 1984;106:405−15.

[13] Kennedy FE. Frictional heating and contact temperatures. In: Bharat Bhushan, editor. Modern tribology handbook, vol. 1. Boca Raton, Florida: CRC Press; 2001. p. 247 [chapter 6].

[14] Kerridge M. Metal transfer and wear process. Proc Phys Soc B 1955;68(7):400−7.

[15] Jacobs TDB, Carpick RW. Nanoscale wear as a stress-assisted chemical reaction. Nat Nanotechnol 2013;8:108−12.

[16] Welsh NC. The unlubricated sliding of steel. Phil Trans R Soc Ser A 1965;257:31−61.

[17] Gerberich WW, Cordill MJ. Physics of adhesion. Rep Prog Phys 2006;69:2157−203.

[18] Adams JB, Hector Jr LG, Siegel DJ, Yu H, Zhong J. Adhesion, lubrication and wear on the atomic scale. Surf Interface Anal 2001;31:619−26.

[19] Archard JF. Contact and rubbing of flat surfaces. J Appl Phys 1953;24(8):981−8.

[20] Kayaba T, Kato K. The adhesive transfer of the slip-tongue and wedge. ASLE Trans 1981;24(2):164−74.

[21] Finkin EF. An explanation of the wear of metals. Wear 1978;47(1):107−17.

[22] Rigney DA. Comments on the sliding wear of metals. Trib Int 1997;30(5):361−7.

[23] Godet M. The third-body approach: a mechanical view of wear. Wear 1984;100 (1−3):437−52.

[24] Chiou YC, Kato K. Wear mode of micro-cutting in dry sliding friction between steel pairs (part I): effect of attack angle of the specimen. J Jpn Soc Lubr Eng. 1988;9:11.

[25] Khruschev MM, Babichev MA. Research on wear of metals. East Kilbridge: National Engineering Laboratory; 1966, Chapter 8, NEL Translation 893.

[26] Sundararajan G. The differential effect of the hardness of metallic materials on their erosion and abrasion resistance. Wear 1993;162-164(Part B):773−81.

[27] Sundararajan G. A new model for two-body abrasive wear based on the localisation of plastic deformation. Wear 1987;117(1):1−35.

[28] Kato K. Abrasive wear of metals. Trib Int 1997;30(5):333−8.

[29] Archard JF. Wear theory and mechanisms. In: Peterson MB, Winer WO, editors. Wear control handbook. New York: ASME; 1980. p. 35−80.

[30] Lee GY, Dharan CKH, Ritchie RO. A physically-based abrasive wear model for composite materials. Wear 2002;252(3-4):322−31.

[31] Karmakar S, Rao URK, Sethuramiah A. An approach towards fatigue wear modeling. Wear 1996;198(1-2):242−50.

[32] Laine E, Olver AV, Lekstrom MF, Shollock BA, Beveridge TA, Hua DY. The effect of friction modifier on micropitting. Trib Trans STLE 2009;52(4):526−33.

[33] Tallian TE. Simplified contact fatigue life prediction model—part II: new model. J Tribol ASME 1992;114(2):214−20.

CHAPTER 4

Boundary Lubrication Mechanisms and Modeling

4.1 INTRODUCTION

Lubricated contacts are far more common than dry contacts, and control of wear and failure in these systems is important. As discussed in Chapter 1, when the load is supported by the asperities via molecular films the lubrication regime is referred to as boundary lubrication. The behavior of these films is altogether different from fluid films and is governed by adsorption. These films, when ideally effective, totally prevent metallic contact and there will be no wear. The friction is due to the shearing of the adsorbed molecular layers. Adsorption is stronger when polar molecules are involved and is an effective way to improve boundary lubrication. Polar additives like fatty acids and alcohols in small percentages in mineral oil are used for this purpose.

Conventionally adsorption-based lubrication is referred to as boundary lubrication. In this chapter it is defined in a broader sense, and refers to lubrication with adsorbed and/or reacted films. When operational severity increases the adsorbed films are no longer effective. Chemical additives are used in such cases to protect surfaces and control wear. Such additives react with the surfaces chemically, resulting in a reaction film.

In this chapter adsorption-based and reaction-based boundary lubrication are considered in two major sections. In the first section the early models are described in a consolidated manner for monolayers and dynamic systems. A consideration is then given to friction and wear. The lubricant failure leading to scuffing is the next topic considered. This is followed by a consideration of lubricant tribology at the nanolevel and its applications. Also an attempt has been made to relate the macro- and nanolevel lubrication. The second major section deals with chemical additives and their mechanism of action. The upcoming area of lubrication with nanoparticles is included at the end of this section. The final section deals with the evaluation methodologies and their limitations. The words macro, micro, and nano in this context need explanation. In the context of this book these define the size of apparent contact.

Modeling of Chemical Wear.
DOI: http://dx.doi.org/10.1016/B978-0-12-804533-6.00004-4
© 2016 Elsevier Inc.
All rights reserved.

Though a contact is macro the real areas of contact are usually micro-sized. In a micro contact the real spots may be nano-sized. If the size is nano the real contacts can be subnano or atomic.

4.2 ADSORPTION BASED BOUNDARY LUBRICATION

4.2.1 Monolayers

Physical adsorption of molecules on surfaces is due to van der Waals forces and is reversible. The tenacity with which the molecules are adsorbed is related to the heat of adsorption. It is known that hydrocarbons are adsorbed weakly and most of the work in the area of boundary lubrication is conducted with polar molecules that display stronger adsorption. Polar molecules are those molecules that have a charge separation and leads to stronger adsorption via the polar head of the molecule. For example, the (OH), (NH$_2$), and (COOH) groups in alcohols, amines, and fatty acids are responsible for the stronger interaction. Yet another issue of importance is the possibility of chemisorption on reactive metals. When chemisorption occurs there is a covalent bonding at the interface and this bond is much stronger than physical adsorption. Chemisorption is restricted to one monolayer. For example, when a monolayer of stearic acid is adsorbed on a copper surface in normal atmosphere a monolayer of copper stearate may form due to chemisorption. The presence of an oxide layer on a surface may be necessary for the chemisorption to occur. Chemisorption is also treated under physical adsorption. The heat of adsorption in this case is significantly higher than in physical adsorption. Figure 4.1 distinguishes adsorption and chemisorption.

The fundamental studies were conducted with preformed monolayers on different metals. Early investigations were due to Hardy [1], Bowden and Tabor [2], Zisman [3], Cameron [4], and others. The studies were conducted in friction machines with different materials made as smooth as possible with a pin on a flat configuration. The speeds involved were less than 10 mm/s while the loads were less than 100 g. In ideal sliding the friction occurs between monolayers with no metal contact.

The ideal situation is illustrated in Figure 4.2.

In most of these tests, although some asperity contact was involved it was very low and the frictional behavior can be attributed to boundary films. Friction force, durability of films, and temperature-related failure detected by a sudden transition of friction formed the basis for understanding the film behavior.

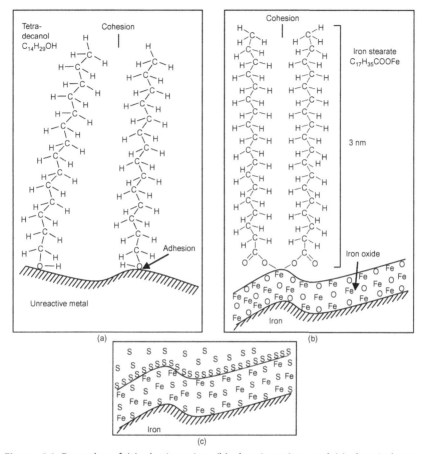

Figure 4.1 Examples of (a) physisorption, (b) chemisorption, and (c) chemical reaction on a steel surface due to EP additive.

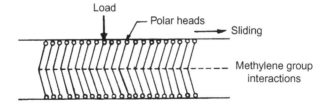

Figure 4.2 Interacting boundary layers with flat surfaces.

Table 4.1 Influence of chemisorption on friction

Surfaces	Paraffin oil	Paraffin oil + 1% dodecanoic acid
Nickel	0.3	0.28
Chromium	0.3	0.3
Platinum	0.28	0.25
Silver	0.8	0.7
Copper	0.3	0.08
Cadmium	0.45	0.05
Zinc	0.2	0.04
Magnesium	0.5	0.08

The major findings still valid today are summarized as follows:

- The more polar an additive the better the lubrication since heat of adsorption is higher with more polar compounds.
- In a given series higher molecular weight compounds have better durability and frictional behavior in view of increased cohesion between molecules.
- Lubricant failure is related to a critical temperature that depends on adsorption/chemisorption.
- Under stress the adsorbed layers bend significantly, and this has a bearing on the friction at the interface.

Some of these ideas were first demonstrated by Hardy [1]. The early studies by Bowden and Tabor [2] have shown the difference between physisorption and chemisorption in boundary lubrication. When tests were done with paraffin there was high friction with f values ranging from 0.3 to 0.8 for all metals tested. When 1% by weight of dodecanoic acid (lauric acid) was added the friction for unreactive metals is similar whereas for reactive metals the f values reduce drastically to 0.04−0.08 due to chemisorption, resulting in a soap monolayer. This is illustrated in Table 4.1.

Some other early studies showed that in the case of soap formation failure occurs at the melting point of the soap. These early fundamental ideas gave strong impetus for further studies in the area.

4.2.2 Real Systems and Dynamic Adsorption

The life of monolayers is limited. While early work indicated that multilayers improved durability [4], the life of such layers falls far short of the practical requirement. In real systems the layer needs constant renewal as it wears out. This is achieved by incorporating the polar compound as an additive in the base fluid. A detailed coverage of the modeling based on dynamic adsorption is available in Chapter 4 of LWST. Here only a summary of the available ideas is given:

1. The strength of adsorbed layers is related to heat of adsorption.
2. Wear in boundary contacts depends on the extent of a metallic contact termed as fractional film defect.
3. Increasing contact temperature reduces film coverage and finally leads to lubrication failure.
4. Models are inadequate to quantitatively predict wear or lubricant failure but are useful in relative selection of polar additives.

There are several aspects on which clarity is not available and research is being conducted. The major investigations being conducted are related to the following:

- The role of mono- versus multilayers in boundary lubrication that will be brought up later in the chapter.
- The complex role of surface oxides and hydroxides on adsorption/ chemisorption that depends on acid−base interactions, catalytic effect of nascent metals, and exo-electron emission [5,6].

4.2.3 Friction and Wear in Boundary Contacts

4.2.3.1 Friction

The early observations on friction were due to Hardy [1]. This work was conducted with excess lubricant either in solution form or as it is. It is assumed that the lubrication behavior is due to the adsorbed monolayers on the surfaces. This assumption is justified by a large body of later work. His historical work is shown in Figure 4.3. The figure shows that friction coefficients decreased with molecular weight. This has been attributed to effective packing of molecules. For acids μ saturated at a molecular weight of 200. More detailed work by Zisman [3] with monolayers on glass and steel surfaces confirmed the trends for acids and alcohols. With alcohols the leveling off occurred at a molecular weight of 280 with μ value around 0.05. For fatty acids the μ value saturated at a molecular weight of 200 to around 0.05. The points of saturation were attributed to formation of condensed layers. This was verified indirectly by contact angle measurements.

More sophisticated work has directly characterized the monolayers with regard to structure as well as molecular density [7] and studied friction and durability. This work forms an important link with the older work.

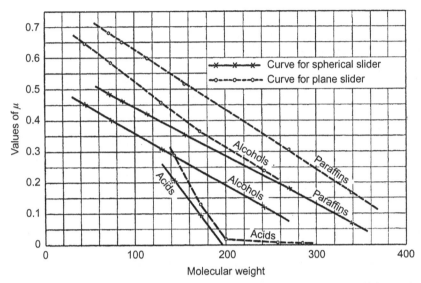

Figure 4.3 Influence of molecular weight on boundary friction. *Reproduced from [1].*

The friction behavior was similar to that observed in earlier literature though actual values vary between steel and glass surfaces. Another important aspect reported in this work is the durability of films. For fatty acids it was shown that durability of films increases with molecular weight in the range of 18−26 carbon atoms per molecule though friction coefficient saturates at 18 carbon number. The durability is higher with glass surfaces as compared to steel. Also the friction was lower for glass surfaces. Thus base material also has an influence on boundary lubrication as expected. Similar findings were made by Zisman [8].

It is now agreed that low friction is due to condensed solid-like films. There is further research and understanding in this area with atomically smooth surfaces that will be discussed next. Another aspect of interest first identified by Bowden is the influence of pressure on the monolayers. The friction coefficient was found to be independent of pressure and it was argued that shear strength increases linearly with pressure as originally proposed by Bridgman [9].

Reference also needs to be made here to the careful studies conducted with preformed mono- and multilayers by Briscoe [10]. The studies include the influence of pressure, shear rate, and temperature on friction. These studies directionally are similar to earlier studies but show that the detailed behavior is system-specific. Also it was reported that with increasing temperature the friction decreases. A thorough review of the earlier work was reported by Campbell [11]. Friction in dynamic situations is difficult to model and current evaluations are made mainly in test rigs under different operating conditions. For example, selection of friction modifiers can be based on their relative performance in different rigs.

4.2.3.2 Wear

Surfaces are not ideally smooth and asperity contacts are highly stressed. The real contact area A_r in boundary lubrication is the same as in the dry case except that boundary layers separate the real areas. The highly stressed film will have defects where metal contact can occur. The metal contact is characterized by fractional film defect, α, which is defined as the fraction of real area that is metallic. The word metallic needs to be qualified. The surfaces are usually covered with oxides and the contacts may as well be a mix of metallic and oxide contacts. The contact situation with defects is illustrated in Figure 4.4.

If shear strength of the film is s_l and s_m is the shear strength of the metallic interface it can be shown that the friction force F is

$$F = A_r[\alpha s_m + (1 - \alpha)s_l] \qquad (4.1)$$

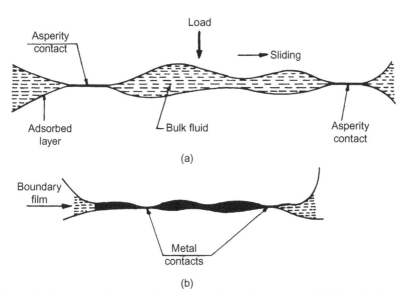

Figure 4.4 (a) Interacting boundary layers with real surfaces and (b) detail of metal contact at asperity due to film defect.

It follows that friction coefficient f can be expressed as

$$f = \frac{F}{W} = \alpha f_m + (1 - \alpha)f_l \tag{4.2}$$

where

W = load,

f_m = friction coefficient of the metallic interface,

f_l = friction coefficient of the boundary film.

The values of f_m and f_l can be measured from dry contact and defect-free boundary contact. Also f can be measured during the experiment. This will permit calculation of α from the above equation. However the zone of interest is for low values of α that leads to small variations in f that are difficult to measure. Also assigning specific value for f_m based on dry contact is questionable. The value of f_l can also change with operating conditions. Hence this approach is not feasible to obtain α value.

An equation based on fundamental considerations was proposed by Rowe [12] in the 1960s to obtain metal contact. This equation is based on the mean time of stay of a molecule and the time needed to move the surfaces one molecular diameter. Without going into the details of derivation the final equation for α is

$$\alpha = \frac{x}{ut_0}e^{-E/RT_s} \tag{4.3}$$

where

E = heat of adsorption (J/mol),

T_s = temperature of the surface (K),

R = molar gas constant,

x = diameter associated with an adsorbed molecule (m),

u = sliding velocity (m/s),

t_0 = fundamental time of vibration of the molecule (s).

The expression for α is applicable for values α of less than 0.01. A more complex equation is available for larger values of α. The wear rate can now be expressed as

$$\frac{V}{l} = K\alpha\frac{W}{H} \tag{4.4}$$

This equation is based on Archard's equation for adhesive wear derived in Chapter 3. It is assumed that wear coefficient of the metallic contact is governed by the dry wear coefficient K in the zone of fractional metal contact. The wear rate can also be expressed as

$$\frac{V}{l} = K_b \frac{W}{H}$$ (4.5)

where K_b is called boundary wear coefficient. As $K_b = K\alpha$, it varies significantly with operating conditions. This is because α varies exponentially with temperature and with sliding speed.

This wear model has been extensively studied [13,14]. Quantitative agreement with the model was inadequate. There are several reasons for this that have been elaborated in Chapter 5 of LWST. First, heat of adsorption on an active surface can be very different from static test values. Second, it is difficult to get values for t_0 with certainty even for a simple hydrocarbon molecule. Additionally the estimation of contact temperature can vary depending on the approach used. In particular the experimental studies showed that influence of temperature on α is clearly less than exponential. The problem of adsorption versus chemisorption for different surfaces further complicates modeling.

One important aspect not considered in the above modeling needs elaboration. This equation implies that wear rate with a given K is simply obtained by multiplying with *fractional area*. However the wear rate is based on volume. If it is assumed that each metal contact has on average an area equal to the asperity contact area multiplied by α then the diameter of the metal contact spot will be $d\alpha^{1/2}$. The wear volume generated over a sliding distance d and following the derivation for adhesive wear of Archard it can be shown that wear rate equation now becomes

$$\frac{V}{l} = K\alpha^{3/2} \frac{W}{H}$$ (4.6)

The actual contacts are a mix of metallic and oxides and it is assumed that K has a specific value for such contacts and is obtained experimentally. This concept is similar to that invoked in Chapter 3 where it was used in relation to the probability factor.

This modification makes a large difference. An example may illustrate this. From [12] the wear rate observed at 8 kg load was 2×10^{-8} cm^3/cm. With the given values in this paper α value will be 6.7×10^{-5}. If the modified equation is used for the same wear rate α will be 1.6×10^{-3}. The modified equation shows that for a given α the wear rate will be less by a factor of $\alpha^{1/2}$ and this change will be substantial. While it is known that wear particles will have a distribution this can only mean a modification in the equation. The wear now represents the cumulative effect of a range of α values.

The interest in modeling boundary wear has waned partly due to the difficulties just pointed out. The interest may revive because of the large scale use of eco-friendly fatty oil-based materials in the future. Also wear in the presence of self-assembled monolayers (SAMs) is assuming importance in MEMS and modeling will be required. It is hence necessary to take into account the size effects discussed here while modeling.

4.2.4 Scuffing Phenomena and Modeling

Wear in lubricated contacts increases with severity of operation as expected. When a critical condition is reached the lubricant fails and the total load is borne by the asperities. This sudden change in many cases leads to catastrophic failure of surfaces called scuffing. The definition of scuffing by the Institution of Engineers [15] is "Gross damage characterised by the formation of local welds between surfaces." Another definition due to OECD [16] is "Localised damage caused by the occurrence of solid phase welding between scuffing surfaces without local melting." Ludema [17] considered scuffing to be due to roughening of surfaces by plastic flow with or without material transfer. Scuffing is a complex process leading to failure and depends on the nature of materials and operating conditions. The appearance of scuffed surfaces depends at what stage of the scuffing process the machine is stopped and surfaces examined. This may be one of the reasons for varied definitions. From a practical point of view it may be defined as a surface damage that amounts to failure. In some cases surfaces can even seize. In a test machine like 4-Ball, the seizure with mineral oil alone at relatively low loads can be demonstrated.

No tribological contact can be allowed to scuff. As mineral oils are normally used the problem of scuffing was investigated in varied machines. The first generalization attempted was due to Blok [18] in the 1930s. He proposed that mineral oils fail at a critical contact temperature, T_s of 150 °C. He proposed an equation to find temperature rise (flash temperature) in the contact due to frictional heating. The contact temperature is defined as

$$T_c = T_b + T_f \tag{4.7}$$

where T_c is contact temperature, T_b is bulk temperature, and T_f is flash temperature. The concept involved is that lubricant fails at a critical temperature resulting in scuffing. A large amount of research was done on this aspect and varying views were expressed on this issue. It is now accepted that failure temperature need not be constant and can vary with operating conditions. The present status of this idea is its use as a design guide to prevent scuffing in gears. The procedure is described in AGMA 2001-B88. This approach is based on the work of Enrrichello [19]. In this approach limiting temperatures are defined as a function of oil viscosity. The design is considered safe if the estimated T_c is less than T_s. The recommended T_s values in F are given by the following equations:

$$T_s = 146 + 59 \ln(\nu_{40}) \text{ for mineral oils without anti-scuff additives} \quad (4.8a)$$

$$T_s = 245 + 59 \ln(\nu_{40}) \text{ for the oils containing anti-scuff additives} \quad (4.8b)$$

While the earlier equations were based on constant T_s, in the present equations critical temperature depends on viscosity. This is because gears operate in EHL regime and transition to boundary lubrication, and scuffing is related to viscosity. Thus higher viscosity fluids should have higher failure temperature.

Evidence from laboratory work under boundary-lubricated conditions also suggests that mineral oils as well as polar compounds fail below 150 °C. The monolayer studies indicate lower failure temperatures based on melting temperatures of soaps or desorption as the case may be. In dynamic situations where the additives are dissolved in base oils the failure temperatures can be different. Even with mineral oils alone failure temperatures around 150 °C are observed by many workers [20,21]. This can be due to impurities of polar nature in mineral oils. Note that most of these observations pertain to steel surfaces. While the situation is somewhat vague the overall consensus is that mineral oils as well as polar compounds fail to lubricate around 150 °C. However the chemisorbed layers like soaps are very beneficial in terms of friction and wear at lower temperatures.

The large body of experimental work done so far is unable to clarify the scuffing mechanism(s) and it is not intended to get into the details of these studies. Taking into account the experimental observations on scuffing the authors consider that scuffing is due to sudden *transient* growth in contact area and depends on critical metal contact level through oxides.

The sequence of events leading to scuffing can be envisaged as follows. As severity increases direct asperity contact increases. The asperities are covered with oxide films and the direct metal contacts occur through the films. Let this be expressed as fractional defect β within the oxide film. With increased β adhesive wear increases, resulting in larger wear particles due to increased metallic content. But the oxide and adsorbed lubricant are able to prevent contact growth. When metal contact reaches a critical level the adhesive junctions grow, leading to plastic deformation and scuffing. It is considered that this junction growth, however short, is similar to the case of contact growth in ultra high vacuum and is governed by combined stresses. But soon enough the runaway situation increases temperature and oxidation re-establishes and controls further growth. However the damage is already done and the components are no longer functional. When lubrication failure occurs at a given temperature the total load shifts to asperities and critical β may occur at this point or beyond. But for materials that have brittle surface oxide like aluminum alloys, the scuffing loads are known to be lower than steel. This is due to the fact that critical metal contact can occur easily through brittle oxide films. These failure temperatures can be below lubricant failure temperature. This hypothesis seems to explain the major experimental findings as given here:

- Earlier literature suggested that increase in dissolved oxygen increases the scuffing load. Evidence also exists that adhesive wear due to metal contact is reduced when oxygen content in the lubricating oil increases [22].
- In any scuffing experiment despite sudden increase in friction it tends to stabilize at a high value. This is because the high temperatures once again provide adequate protection and prevent growth in contact area due to oxidation.
- The model can reconcile with plastic deformation without invoking high temperature plasticity.
- According to this hypothesis near-surface plastic deformation and high friction is a *consequence* of short-lived contact area growth. This idea differs from a recent proposal that scuffing has its origin in the subsurface plastic deformation [23].
- Some work in modest vacuum of $10^{-4}-10^{-6}$ Pa at low temperatures of $100-250$ K has evidence of scuffing due to growth in contact area as the oxide is removed [24]. Hence high temperature is not a prerequisite for growth in contact area and scuffing.

- It is well known that cold welding is possible between metals provided the surface oxides can get displaced, allowing atomic contact of metals. This also supports the possibility of scuffing at critical conditions involving oxide cracking/displacement.

The hypothesis is plausible but cannot be quantified as the mechanism is complex. One possibility is to consider critical temperature of lubricant failure as a conservative criterion for scuffing of bearing steels. This criterion may not be applicable to other materials like aluminum alloys.

4.2.5 Lubricant Tribology at the Nanolevel

Tribology at the nanolevel is gaining importance as this understanding is needed in micro and nano devices. Nanotribology involves contact over a nanoarea usually between atomically smooth surfaces. The atomically smooth surfaces avoid roughness problems inherent in macroscopic contacts. It was hoped that fundamental understanding of friction is feasible at this level and can even be connected to the macro friction. But the very complex friction including stick-slip has been difficult to model and no generalizations are possible. The investigations involve both dry- and boundary-lubricated contacts. The present section will be confined to boundary-lubricated contacts.

The study of atomic contacts has been made possible by two kinds of apparatus, surface force apparatus (SFA) and atomic force microscope (AFM). The SFA has been extensively used for studying thin liquid films. In this apparatus crossed cylinders result in circular contact. Mica is glued to the surfaces resulting in circular contact at the mica surfaces. Mica is atomically smooth and molecular layers of different types can be studied in this apparatus. Both normal and tangential forces can be measured. The film thickness can be measured by interference techniques. Typical contact pressures involved are in MPa and typical contact areas involved are in the range of $100-200 \, \mu m^2$. Multi- to monolayers are involved in these studies. The apparatus more widely used now is AFM. A thorough review of its use in single-asperity nanotribology is available [25]. The schematic diagram in Figure 4.5 reproduced from [25] shows the operating principle of AFM. The operation can be in different modes and can be adapted to measure adhesive forces by approach/retraction, surface roughness, and friction. When used for measuring friction it is usually referred to as friction force microscope (FFM). In this apparatus a

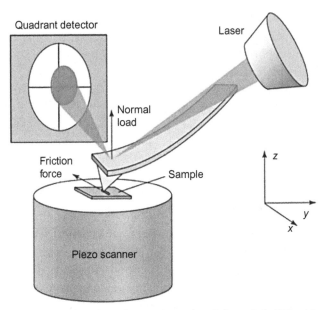

Figure 4.5 Schematic diagram of FFM. *Reproduced from Ref. [25] with permission from IOP Publishing Limited.*

sharp tip of radius ranging from 10 to 100 nm can be brought into the proximity of the lower surface. The cantilever deflects vertically depending on the attractive/repulsive forces acting on the tip. The tip can be moved relative to the stage and the lateral friction forces result in twisting of the cantilever. A laser beam is deflected from the back of the very flexible cantilever onto a four-quadrant position-sensitive photo detector. The position is related to the deflections and is processed to obtain normal and friction force. The resolution levels are usually high and are in the sub-Angstrom range. The contact areas involved are very small and stresses into the GPa range can be applied in this instrument. Interpretation of AFM data is not easy and has uncertainties related to the spring constants, resolution of normal and axial forces, and environment. The commercial cantilevers are made mainly of silicon or silicon carbide.

The behavior of confined molecular layers has been researched with these apparatus and found to be complex. Modeling of the observed behavior by molecular dynamic simulation is also in progress. The major trends observed with these apparatus may be summarized as follows:

1. When any fluid is compressed between atomically smooth surfaces the fluid develops an order once the thickness goes below about 10 molecular layers. This ordering is observed both in SFA and FFM. Also this ordering was not observed with branched chain molecules. As the thickness reduces to monolayer the film behaves like a solid with a crystalline or amorphous structure depending on the nature of surfaces [26].
2. Friction experiments show a far more complex behavior. The boundary film may behave like crystalline solid, amorphous solid, or liquid-like, depending on the operating conditions and film thickness [27].
3. Atomic contacts may involve stick-slip behavior due to altering phase behavior during sliding. This atomic stick-slip is sensitive to shear rate [27].

The major interest in nanolubrication is the behavior of SAMs. These are specially tailored molecules and the main aim is to use them for sliding contacts in MEMS devices. Tribology with different functional groups and molecular organization is being probed [28]. It is interesting to know that broadly chain length effects are similar to those observed in macrotribology.

4.2.5.1 Relevance of Nanolubrication to the Macro Level

It is of interest to know what impact the nanolevel understanding has on the macro level boundary lubrication. At the macro level the actual contacts are micro-sized and the boundary films are mono- or multilayer. Hence it should in principle be possible to relate the two domains.

At the macro level stick-slip is not observed with polar compounds like fatty acids. In fact they are used to obviate stick-slip in applications like machine slide ways. Atomic stick-slip needs a well-defined structure and a very low shear rate. At macro level crystalline ordering may not occur due to roughness effects. Also the shear rates in engineering contacts normally range from 10^6 to 10^8 s^{-1} and is orders of magnitude higher than in atomic friction experiments that may range from 10^3 to 10^4 s^{-1}. Thus a reasonable case may be made to explain absence of stick-slip in macro contacts. But there may be cases of very slow movements in precision engineering and in controls. In such cases stick-slip may be relevant.

The other issue is the nature of friction force observed in the two domains. The significant variations for a given molecular structure observed in FFM as a function of operating conditions are not apparent at the macro level. As stated earlier the shear strength varies directly with pressure to a first approximation. Interesting work reported by Briscoe and Evans in SFA [29] also shows direct dependence of pressure on shear strength. They also reported a *linear* decrease in friction with temperature. The observation related to pressure is similar to the macro observation. This may be because the shear rates involved in SFA are higher than in FFM. Hence it may be said that shear rate may be a major factor in relating macro and nano research.

The detailed modeling at macro level needs knowledge regarding entrainment into asperity contact, squeezing out of the fluid, and the eventual shearing of the boundary layers. Also the question in macro lubrication whether mono- or multilayers are involved remains unanswered. If FFM studies are conducted with a microdrop of lubricant instead of a monolayer this may mimic the macro contact. Such studies may help in further bridging the gap. The earlier work by Spikes [30] is of relevance here with regard to the formation of boundary layers. He studied lubricants in rolling/sliding contacts using a special rig. Ultra thin film interferometry technique was used to find EHL film thickness down to nanolevel. While the surfaces have nano roughness it was assumed that they flatten under pressure and thickness measurement down to nanolevel is feasible. This work showed that for hexadecane the film thickness up to a low value of 1 nm can be reconciled with continuum EHL theory. The polar additives can increase the film thickness at some operating conditions and their molecular thickness can be assessed. The operating conditions here are closer to engineering situations and hence the work forms a link between macro and nanotribology of lubricants. However this work still does not represent pure sliding contact. Some recent work attempts to model nano scale lubrication in EHL using molecular dynamic simulations as continuum theory may not be applicable. Also surface slip is considered in this modeling [31]. It is hoped that future research may effectively bridge the gap between nano and macro level understanding.

4.2.5.2 Applications of Nanotribology
The success of nanotribology in terms of actual applications is limited. The expectation is that the work will be useful in the near future particularly in small devices like MEMS. Stiction is one area where the fundamental research has found application and is described below.

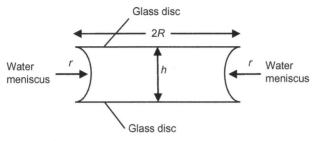

Figure 4.6 Adhesive forces due to meniscus formation.

Stiction is the resistance at start due to strong adhesive forces. The strong adhesive force referred to here is due to capillary forces. Such forces develop when a liquid meniscus forms between the surfaces. Figure 4.6 shows a water meniscus between two circular surfaces.

The equation for force between two surfaces of radius R each separated by a liquid film of height h is

$$L = \frac{\pi R^2 \gamma}{r} \qquad (4.9)$$

where L is force, r is the meniscus radius, and γ is the surface tension of the liquid. The radius of meniscus is given by $r = h/(\cos \alpha + \cos \beta)$ where α and β are the contact angles at the two surfaces.

An example can clarify the large adhesive force. Consider water film of 1 µm between two glass surfaces. The surface tension of water is 73 mNm^{-1}. Assume water perfectly wets glass so that $\alpha = \beta = 0°$. So, $r = h/2 = 0.5$ µm. Adhesive pressure $= L/(\pi R^2) = \gamma/r$. With these data we calculate adhesive pressure equal to 0.146 MPa, which is a large pressure indeed!

It can be now appreciated that meniscus forces in MEMS can be very problematic. An example of the problem in MEMS is illustrated in Figure 4.7 [32]. In micromachining the sacrificial layers are finally removed by a liquid etchant like hydrofluoric acid. Subsequently the acid is rinsed off with water. Then during drying the water meniscus that forms exerts strong attraction which causes the micro-sized surfaces to get stuck, and the component becomes nonfunctional.

It may be clarified that the above is an example of strong adhesive forces and the damage they can cause to the surfaces. Stiction on the other hand refers to the problem associated with the starting torque required to overcome the adhesive forces. This involves the rupture of the water film present as sliding commences. In AFM this rupture is

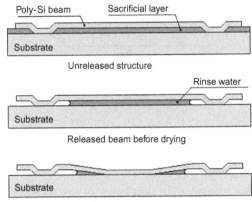

Figure 4.7 Release problem in MEMS components due to meniscus forces. *Attributed to Chollet and Liu [32].*

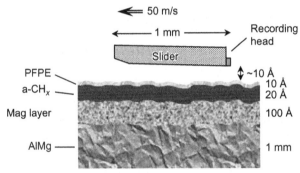

Figure 4.8 Illustration of computer head—disk interface lubrication. *Reproduced from Ref. [33] with permission of John Wiley. This figure is a modern version of that in the reference kindly provided by the author.*

characterized by a break-free distance needed to separate the surfaces during retraction. Stiction was observed in computer head—hard disk interface several years back. The roughness of disc surface is of nanolevel. The interface can develop water meniscus in a humid atmosphere. When the system is started the stiction can cause a crash due to high initial friction. This problem has been overcome by applying a lubricant layer on the top that forms a hydrophobic layer. Such a layer does not allow meniscus formation. The layer also provides lubrication.

The lubrication properties and durability of films have been thoroughly investigated. The nature of lubrication involved has been well described by Gellman [33]. Figure 4.8 is based on the expected future

development and shows the various fine dimensions involved in the lubrication. This figure is provided by Gellman and is a more modern version
of that given in Ref. [33] and represents the future scenario. The disk has
a magnetic layer on the top of which is a diamond-like carbon (DLC)
layer of nano dimensions. The topmost lubricant layer is 1 nm and is a
SAM layer of perfluoropolyether (PFPE). The disk runs at $\sim 10,000$ rpm
and the head flies at a height of $\sim 10\,\text{Å}$ supported on an air bearing.
Such close tolerances can cause rubbing particularly at start-up besides
stiction discussed earlier. The lubricant also protects when there are any
contacts during running. Tailoring lubricants for this purpose may be
considered a major achievement of nanotribology. Present clearances
between disk and head easily exceed 20 Å. Also the carbon and magnetic
layer thickness may be twice the values shown here.

The major development of the SAM layer has been possible because of
detailed studies at the nanolevel with AFM. The lubricant layer currently
used consists of a blend of PFPE molecules with reactive and unreactive
end groups that provide optimum performance in terms of durability and
friction. Such lubricants are also being studied for MEMS applications.

A typical example of study on adhesion is that conducted by Mate [34]
with AFM on silicon surfaces and is reproduced in Figure 4.9. This work
shows the nature of forces observed when a tungsten tip of ~ 100 nm
radius approaches the silicon surface covered with polymer lubricant. Two
types of PFPE lubricants are studied. In one case the end group is unreactive $-CF_3$ and in the other case the end group is reactive $-CH_2OH$. The
point M in the diagram represents sudden adhesive force due to meniscus
formation and the corresponding attractive force. As the tip is brought
closer the unreactive polymer shows typical behavior with attractive van
der Waals coupled with meniscus forces. This is in contrast to the repulsive
forces that act between reactive end groups for the reactive polymer.
Without going into further details, in practice this translates into high resistance to displacement from the surface of the reactive polymer. It means
much higher pressure is needed to squeeze out the reactive layer. Such
studies are now conducted with different molecules for MEMS applications. Durability and friction of such layers are also being explored.

The question that arises is whether SAM layers can be used to control
friction and wear in sliding MEMS components. It is known that silicon
surfaces are very poor with regard to wear and in fact no commercial
components are available that involve sliding. The problem is being
researched from different angles. Surface modifications that involve solid

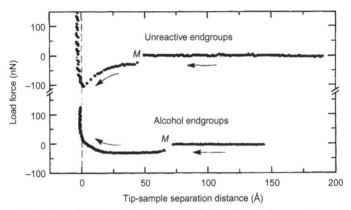

Figure 4.9 Behavior of PFPE lubricant layers with reactive and unreactive end groups in AFM. The nature of forces observed during approach are shown. *Reprinted with permission of Mate [34]. Copyright 1992, by the American Physical Society.*

film coatings such as DLC, ion implantation, and laser texturing are being studied. As silicon friction depends on crystal orientation this possibility is also being studied. The lubricant coating is another approach. This can be done on the modified or original surfaces. This is the area where SAMs are being extensively investigated. The molecules being studied include PFPEs, different silanes, and silicones. With SAM technology the molecules can be tailored with different reactive groups and grafted to surfaces. The major problem in lubrication is that continuous lubricant supply is not practical for the small components. Hence reliance has to be placed on long durability of the SAM layers. This is a very difficult requirement to meet. In the computer application cited above the problem is mainly during start-up and there is no continuous rubbing of the film. This has been met by careful tailoring of SAM layers as already discussed. At present it appears there is no satisfactory answer to the MEMS wear problem via lubricant approach. Some effort is also being made with continuous vapor lubrication, but its future again is uncertain.

Real MEMS systems have nanolevel roughness and in many cases the contact is between silicon surfaces that have an ill-defined oxide layer. It is now increasingly considered that evaluations should be done with actual MEMS components. Tribological evaluation using MEMS components is now prevalent. The types of machines being used are well described by Williams [35]. However the fundamental studies at the molecular level are useful in developing basic knowledge that is necessary in lubricant and material selection.

4.3 BOUNDARY LUBRICATION WITH REACTION FILMS

Mineral oils with or without polar additives can at the most function up to 150 °C, beyond which the lubricant fails leading to scuffing. However the wear may be unacceptably high even earlier. It is common practice to control scuffing with EP additives while wear control is affected by anti-wear (AW) additives.

The EP and AW additives may be broadly classified as more reactive and less reactive additives. In some cases this distinction is not clear cut. Also in multifunctional additives both functions may be incorporated in the same molecule. The additives involved are sulfur compounds, phosphorous compounds, and multifunctional additives that contain both elements. In scuffing control sulfur compounds react immediately as the lubricant fails, forming a reaction film that prevents direct asperity contact and hence scuffing. The AW additives act more gradually, developing films that are more wear resistant than the base fluid. These additives and their action mechanisms are discussed below class wise while another section deals with evaluation methodologies. Chlorine-containing additives that are being phased out are not discussed here. The last subsection covers the upcoming area of lubrication with nanoparticles.

4.3.1 Sulfur Compounds and Their Action Mechanisms

The use of sulfur to improve lubrication is one of the important ideas that finds extensive use today. The present technology is centered around two major classes of sulfur compounds. These are sulfurized olefins and sulfurized oils and esters. The ability to easily manufacture additives with different levels of sulfur is a key factor in the application of these products. The EP function of these additives is controlled by the level of active sulfur. The sulfurized olefins generally have higher active sulfur and can be used for very heavy-duty applications like shock loading in gears. All these sulfur compounds are extensively used in heavy-duty gear oils and metal working. The ester-based additives have the advantage of lower friction though they are less active. Several issues like solubility and thermal stability have to be considered in selecting these additives. Two typical sulfur compounds, sulfurized isobutylene and sulfurized methyl oleate, are illustrated as follows.

Sulfurized isobutylene	$(CH_3)_3C-[S]_x-C(CH_3)_3$
Sulfurized methyl oleate	$CH_3-(CH_2)_7-CH-CH-(CH_2)_7-COOCH_3$

with the two CH groups bridged by S.

Some other compounds like dibenzyl disulfide and alkyl sulfides are also used as EP agents. The reactivity of these compounds is lower than the earlier discussed compounds. The important issue is how to assess the antiscuff property of an additive. The first attempt was based on the C−S bond strength as proposed by Forbes [36]. Dorinson and Ludema [37] also considered the activity of several sulfur compounds with alkyl and aryl groups in terms of bond strength. For example, disulfides have lower bond strength than monosulfides and hence are considered better EP additives. Also the alkyl and aryl groups influence the bond strength. Such information can aid in selecting additives.

Another interesting approach is due to Sakurai and Sato [38]. In this approach chemical reactivity was characterized by static reaction over a hot wire. As the reaction progresses the thickness increases, which was measured by the change in electrical resistance of the wire. Most of the compounds followed parabolic reaction, which can be expressed as

$$\Delta r^2 = kt \tag{4.10}$$

where

Δr = film thickness (m),
k = rate constant (m^2/s),
t = time (s).

An attempt was made to relate the reactivity at 400 °C with the load-carrying capacity as measured in a 4-Ball tester. The correlation was reasonable. The problem with such studies is the fact that static and dynamic reactions can differ considerably. In fact as shown first by Godfrey [39] the films formed in contact contain both sulfur and oxygen. The action mechanism thus involves a combined action of oxygen and sulfur. The film formed is a mixture of oxides, sulfides, and condensed organic and organo-metallic compounds. Two examples of research from earlier work in this area are given in Refs [40,41]. Hence evaluations are based on performance in standard test machines and the products graded. The next section will deal with tribological evaluation methods and their limitations for both EP and AW properties.

It may be observed that even oxidational wear that depends on reaction with atmospheric air is a difficult problem to model. This is because the oxidation reactions in static and dynamic situations differ considerably.

4.3.2 Phosphorous Compounds and Their Action Mechanisms

The phosphorous compounds used are mainly phosphate and phosphite esters. The very common AW additive in this class is tricresyl phosphate, shown as follows.

The phosphate additives may be characterized by the general formula $O = P(OR)_3$ where the molecule may have alkyl, aryl, or mixed groups. The phosphate esters are prone to hydrolysis and problems of thermal stability. It is considered that aryl groups confer better thermal and hydrolytic stability. Several partially substituted hydrogen phosphates are also used as AW additives. Amine phosphates are another class of additives that find AW application and are less reactive than acid phosphates.

The phosphite esters when fully substituted have the structure $P(OR)_3$. Partially substituted hydrogen phosphites are also widely used as AW additives. Besides their AW properties they have the special capability of functioning under high-torque, low-speed conditions.

Yet another type used are phosphonates, which are more stable due to a direct carbon−phosphorous bond and are suitable for high-temperature applications.

The mechanism of action of phosphorous additives is a huge area of interest with significant controversies. The most studied additive in this class is tricresyl phosphate. Initial explanation was based on phosphide formation on the surface. Later studies, especially those due to Godfrey [42], clearly brought out that the mechanism involved formation of metal phosphate. More recent research also suggests the possibility of formation of polyphosphates with some additives [43]. This can be from impurities in the additive or due to hydrolysis from available moisture in the system. The reactivity was related to the molecule. For example, the acidic phosphates are more reactive than neutral phosphates. With regard to EP properties those compounds that can react quickly are advantageous. Thus acid phosphates and phosphites will be superior to neutral phosphites and phosphates. The AW properties depend on the slower formation and retention of the film and hence AW behavior is likely to be opposite to that of good EP behavior. These observations are only directional and can be influenced by the nature of surfaces, alkyl and aryl groups, as well as environment.

From an actual practice point of view the additive combinations have to be carefully tailored for specific applications. An interesting article discusses the trends in EP additives from the manufacturer's point of view [44].

4.3.3 Multifunctional Additives and Their Action Mechanisms

A combination of AW and EP properties can be achieved by a judicious combination of the two additives. In many cases the EP and AW additives are used in combination for specific systems and the combinations used in some cases are packages, and are proprietary to the manufacturer. The criteria include the reactivity levels needed and balanced functioning with no excessive reaction. One good example is the hypoid rear-axle operation. This presents severe conditions that involve scuffing, shock loading, as well as low-speed, high-torque operation. The additive content for such operation may be as high as 5%. The package may be termed multifunctional as it combines two functions. While this is a common practice, multifunctional additives strictly refer to the molecules that incorporate both functions in the same molecule. The multifunctional additives may also have other properties that are useful.

One important and widely used multifunctional additive is zinc dithiophosphate with alkyl/aryl groups (ZDDP). This is extensively used in engine oils and possesses AW, antioxidant, and mild EP properties. It is also commonly used in steam turbine reduction gears as well as in heavy-duty hydraulic systems. The typical structure is given here:

where R represents alkyl/aryl group

The ZDDP is the most studied additive in view of its extensive use in engine oils that form about half of all lubricants in the world. In view of its pollution due to zinc compounds and the poisoning effect of phosphorous on exhaust gas catalyst its use has to be reduced. One approach is to use alternate ashless additives. The other is partial replacement with other additives that can at least reduce the total phosphorous content. With increasing stringency of laws related to land and air pollution this has become an urgent need. Replacement needs detailed understanding of the action mechanisms. This is the main reason for the voluminous research in this area.

The mechanism involved is complex but some generalizations are possible. It is established that glassy polyphosphates containing Zn and Fe cations are formed that protect the surfaces. The polymers in tribofilms have a layered structure with short-chain length products close to the metallic surface while the top layers in the film are long-chain polymers in the case of alkyl phosphates. With aryl phosphates mainly long-chain polymers are involved. The iron content in the film decreases from the surface. The mechanism of formation is not clear and intensive research in the area has been recently reviewed [45]. Adsorption followed by thermal decomposition is suggested by many as the initial step. The later steps may involve ZDDP molecules interacting with hydroperoxides or oxygen forming reactive intermediates. There is also a possibility of hydrolysis. The role of iron oxide is also not clear. The sulfur in the film is confined to near surface zone and is present in the oxidized form. The AW films with alkyl phosphates are heterogeneous and result in micron-sized elevated patches with corresponding valleys. With aryl phosphates the elevated areas appear as streaks. Between the patches the composition of the film is different from the elevated patches. The average film thickness is of the order of 100 nm. It may be pointed out that these observations involve very modern techniques like XPS, XANES, and AES for surface analysis [45]. Some observations are also based on AFM techniques.

Significant effort is on with regard to mechanical characterization of films by nanoindentation techniques [46]. The upper parts of the film can have hardness values as high as 2−3 GPa and modulus values in the range of 70−100 GPa. Such films may as well be compared to surface coatings used for wear reduction except that they are formed *in situ* and easily replenished (Figure 4.10).

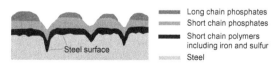

Figure 4.10 Schematic of ZDDP AW film structure.

As ZDDP is used in conjunction with other additives like detergents in engine oils several studies are devoted to the influence of detergents and other additives [45]. The performance in synthetic base stocks is also being studied [47]. Synergy with two important molybdenum compounds, dithiocarbamate and dithiophosphate, is also being intensively explored [48] as the combinations can reduce the phosphorous content for the same performance in engine oils.

Some fundamental studies at the micro and nanolevel with and without ultra high vacuum are providing significant insights into the functioning of chemical additives and adsorbed molecules. These may offer answers to observations that are contrary to the conventional chemistry. For example, the observation of Spikes [49] that ZDDP can form a tribofilm at normal room temperature needs a separate explanation other than decomposition. Work is also being conducted to relate nature of films to the observed wear [50]. These important developments will be taken into account in Chapter 5, which deals with the theoretical modeling of chemical wear.

4.3.4 Nanoparticles as AW/EP Additives

Nanoparticles with dimensions less than 100 nm are being extensively tried for lubricant applications. These products are mainly inorganic and expected to function by interacting with the surface providing AW/EP functions. The goal is to obtain good functioning with a small quantity of nanoparticles by themselves or by partial replacement of conventional additives containing sulfur and phosphorous. The effort is aimed to reduce the conventional additives that have environmental problems. As discussed in the previous subsection, ZDDP in engine oils is being reduced as there is a limit on total phosphorous content. This limit may go down further in the future. In such cases one of the solutions is the use of nanoparticles.

A large number of nanoparticles are being explored. These include copper, copper oxides, silver, MoS_2, molybdenum oxides, titanium oxides, and boron compounds. Several research papers [51−53] are evidence to the growing interest in the area. The bare nanoparticles have high surface energy and tend to agglomerate. These particles can be kept in suspension by capping them with suitable surfactants. The surfactant molecules surround the particle with polar ends attached to the particle. These particles in turn form a stable suspension in oil. Systems with dispersion in aqueous systems are also possible by changing the nature of surfactants. These particles with the surrounding molecules get adsorbed on the surface. How the core nanoparticle interacts with the surface and forms an effective film is a major area of investigation. Going by the results boric acid dispersions and carbon-based dispersions are likely be industrially successful for engine oils and other systems [54]. These developments meant several years of research along with tribological investigations both at bench scale and in actual equipment taking into account the influence of other additives present in the system.

Several scientific investigations at the laboratory level are involved with different levels of sophistication. But in so far as tribology goes many investigators examine the behavior in a model base oil and are not going beyond that. But the vast literature generated in this manner has a potential with regard to preliminary selection on nanoparticle systems. Some interesting research in this area [55] in which Rajesh Kumar is involved is briefly described below as a typical example. It is known that several metals and oxides can be used as nanoparticles. The authors have chosen ceramic nanoparticles (CCZTO) synthesized earlier for microelectronic application. The composition of the nanoparticle is $CaCu_{2.9}Zn_{0.1}Ti_4O_{12}$ and contains calcium, copper, zinc, and titanium. Such particles may generate mixed metallic films in tribological conditions that may be advantageous. These particles sintered for 6, 8, and 12 h have been chosen. The particles were then treated with stearic acid as described in the paper and tested for wear. The 12-h sintered product was not successful, probably due to large size. The TEM image in Figure 4.11 shows the size of particles obtained with 6-h sintering and is ~ 60 nm.

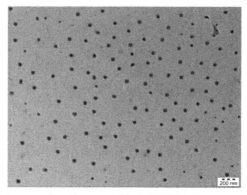

Figure 4.11 TEM image of surface modified SCCZTO-6h nanoparticle of ~ 60 nm size. *Reproduced from Ref. [55] by permission of The Royal Society of Chemistry.*

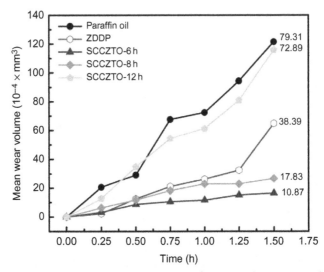

Figure 4.12 Wear comparison on the basis of wear volume. *Reproduced from Ref. [55] by permission of The Royal Society of Chemistry.*

The wear behavior of the particles at 1% concentration shows good wear reduction as compared with liquid paraffin. Also the wear has been analyzed in terms of wear volume and separation of running-in and steady wear zones. This is a better approach to wear instead of simple comparison on the basis of scar diameters. Figure 4.12 shows the comparison. In this figure the overall wear rate on the basis of final wear volume after 90 min test duration is shown at the end of curve.

On the basis of XPS analysis of the films the film contains metal oxides and stearate. The authors proposed that the mechanism involved is tribosintering in the rubbing contact. The sintered film is responsible for wear reduction.

4.4 EVALUATION METHODOLOGIES

The goal of chemical additives is to improve wear and scuffing resistance of an oil formulation. While molecular structure can be related to performance qualitatively it is inadequate to quantify the behavior. So assessments are done in tribological machines that can measure friction and wear. LWST considers the methodologies in detail in Chapter 7. The present section is based on this chapter and limited in coverage.

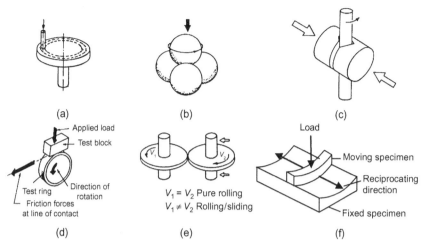

Figure 4.13 Tribological test machine configurations: (a) pin-on-disc, (b) 4-Ball, (c) pin and V-block, (d) block-on-ring, (e) disc—disc, and (f) reciprocating tester.

The section considers the types of test machines, wear evaluation, and evaluation of load-carrying capacity (antiscuff ability).

4.4.1 Types of Test Machines

The machines used with unidirectional testing include pin-on-flat, 4-Ball, block-on-ring, and pin and V-block. These machines mainly operate in boundary lubrication regime. Disc machines and gear machines involve rolling/sliding contact and operate in mixed lubrication regime. Several rigs with reciprocating motion are also available and are used to evaluate ring-liner contacts. These machines are sketched in Figure 4.13.

4.4.2 Wear Evaluation and Its Limitations

Wear can be determined in any machine. However for comparison and laying down specification limits standard test procedures using well-designed machines are necessary. ASTM, Institute of Petroleum (IP), and several other organizations have evolved procedures with which evaluations are done. Procedures are available using 4-Ball, V-block, block-on-ring, and reciprocating testers. The methodology of testing will be exemplified with a 4-Ball machine that is commonly used.

The tests are conducted with AISI 52100 steel balls of 12.7 mm arranged in tetrahedral geometry. The procedure is as per ASTM D 4172-94. In this procedure the tests are run with lower stationary balls immersed in the lubricant. The rotational speed is 1200 ± 60 rpm and the test duration is 60 ± 1 min. The average scar of the three stationary balls is the criteria for wear. As per the standard the repeatability of these tests is 0.12 mm with a 95% confidence level. These organizations have the standard procedures. The limiting values are the concern of those who lay down lubricant specifications. Although the test is simple the interpretation has problems. In reality wear volume and wear rate should be the criteria and not the scar diameter. The wear volume is obtained assuming elastic modulus $E = 208 \times 10^3$ N/mm^2 and Poisson ratio $\nu = 0.3$. The resulting equation that takes elastic recovery into account is given below. Here wear volume V is in mm^3, d is wear scar diameter in mm, and L is machine load in kg.

$$V = [1.55 \times 10^{-2}d^3 - 1.03 \times 10^{-5}L]d \qquad (4.11)$$

Also nondimensional wear coefficient K can be calculated from the following equation.

$$K = \left[\frac{VH}{(23.3Nt)(0.408L)}\right] \qquad (4.12)$$

where H is hardness in kg/mm^2, N is rotational speed in rpm, and t is test duration in min.

The above equations were originally proposed by Rowe [56] and extensively used. More exact calculations show that in Eqn (4.12) the first constant should be 23.035 instead of 23.3. This difference is ignored in the present treatment.

For a selected number of wear scar diameters the wear volume and wear coefficients are tabulated in Table 4.2.

This table shows that for wear scar variation of the order of 0.1 mm the wear coefficient and hence wear rate vary by about 2–8 times. These variations are higher for smaller scars because influence of elastic recovery

Table 4.2 Wear coefficient calculation for some scar diameters

Wear scar diameters, mm	Wear volume, mm$^3 \times 10^3$	Wear coefficient, $\times 10^8$
0.336	0.0591	0.155
0.44	0.400	1.052
0.54	1.100	2.693
0.61	1.895	4.984

is higher for smaller diameters. Thus evaluations with scar diameter alone with repeatability of 0.1 mm can be way off from the reality.

In the above argument it is assumed that final wear rate can be obtained from final scar converted to volume. This amounts to an assumption that wear rate is constant in the test. This introduces an error because the wear rate during running-in and later are different.

Similar problems exist with other testing machines like block-on-ring following ASTM D 2714. Also observations with regard to wear in many research papers do not take into account the running-in aspect. This can lead to wrong conclusions about wear performance. These issues will be considered in detail in Chapter 7 where systematic approaches are described, and is an important part of this book.

4.4.3 Evaluation of EP Properties

These tests refer to the load-carrying capacity of lubricants. The higher the load-carrying capacity the better the seizure protection. As discussed earlier scuffing normally occurs after the lubricant fails. The EP additive offers protection by reacting fast enough to form a reaction film. This film protects the surfaces from scuffing.

The schematic diagram in Figure 4.14 illustrates the action. The additive acts at a threshold temperature and protects surfaces from seizure. It can be seen that the protection extends to high temperature unlike mineral oil, which fails at low temperature. A fatty oil has intermediate

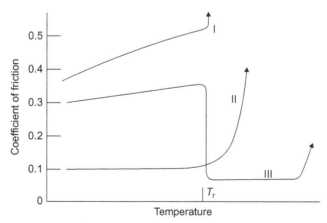

Figure 4.14 EP action of additive (III) with threshold reaction temperature T_r as compared with base oil (I) and fatty oil additive (II). Arrow represents scuffing.

Table 4.3 Test methods for EP properties

ASTM D 2782: Measurement of EP properties of lubricating fluids
(Timkin method)
ASTM D 2783: Measurement of EP properties of lubricating fluids
(4-Ball methods)
ASTM D 3233: Measurement of EP properties of lubricating fluids
(Falex pin and Vee block methods)
ASTM D 5182: Load-carrying capacity tests for oils−FZG machine

behavior. This also means higher load-bearing capacity as higher loading leads to higher temperatures.

Rate processes are involved in protection though they are ill-defined. The basic methodology involves testing at increasing load until failure occurs. Higher loads involve faster progression of scuffing and protection depends on stronger EP action. Finally at a particular load EP film cannot protect and scuffing occurs. The higher the scuffing load the better the protection.

The various procedures involve 4-Ball, block-on-ring, and pin and V-block. In the block-on-ring test the loading is progressive with the same surfaces. In the other two tests fresh test pieces are used at each load stage. A gear machine test based on an FZG gear rig following German standards is also available and is considered to be more relevant for gear oils. The ASTM test numbers are given in Table 4.3 and the detailed procedure in each case is available in the standards.

The methods used are elaborate and are not discussed here. Interested readers may consult LWST Chapter 7, Section 7.4. While the test procedures raise several questions as in the case of wear testing, they are not considered in this book; our focus will be on chemical wear.

NOMENCLATURE

A_r real area of contact
d average asperity contact diameter
d wear scar diameter
E heat of adsorption
E elastic modulus
f coefficient of friction
f_l friction coefficient for boundary film
f_m friction coefficient for metallic junctions
F friction force

h	liquid film height as shown in Figure 4.6
H	hardness of the softer material
k	rate constant, m^2/s
K	wear coefficient, nondimensional number
K	adhesive wear coefficient
K_b	boundary wear coefficient
l	sliding distance
L	adhesive force due to meniscus formation
L	machine load in 4-Ball machine
N	rotational speed in rpm in 4-Ball machine
r	radius of meniscus
R	molar gas constant
R	radius of glass discs shown in Figure 4.6
s_l	shear strength of the film
s_m	shear strength of the metallic junctions
t	time in forming a reaction film of film thickness Δr
t	test duration in 4-Ball machine
t_0	fundamental time of vibration of the molecule
T_b	bulk temperature
T_c	contact temperature
T_f	flash temperature rise
T_s	critical contact temperature
u	sliding velocity
V	wear volume
W	normal load
x	diameter associated with an adsorbed molecule

Greek Letters

α	fractional film defect, fractional metal contact area within real area
α, β	contact angles at the two surfaces in meniscus
β	fractional defect within the oxide film
Δr	film thickness
γ	surface tension of the liquid
μ	coefficient of friction
ν	Poisson ratio
ν_{40}	kinematic viscosity at 40 °C

REFERENCES

[1] Hardy W, Bircumshaw I. Boundary lubrication. Plane surfaces and the limitations of Amontons' law. Proc R Soc Lond A 1925;108:1–27.
[2] Bowden FP, Tabor D. The friction and lubrication of solids, part I. London: Oxford University Press; 1950.
[3] Levine O, Zisman WA. Physical properties of monolayers adsorbed at the solid–air interface-I. Friction and wettability of aliphatic polar compounds and effects of halogenation. J Phys Chem 1957;61(8):1068–77.

[4] Grew WJS, Cameron A. Thermodynamics of boundary lubrication and scuffing. Proc R Soc Lond Ser A 1972;327:47−59.

[5] Mori S, Imaizumi Y. Adsorption of model compounds of lubricant on nascent surfaces of mild and stainless steel under dynamic conditions. Trib Trans STLE 1988;31(4):449−53.

[6] Nakayama K, Leiva JA, Enomoto Y. Chemi-emission of electrons from metal surfaces in the cutting process due to metal/gas interactions. Trib Int 1995; 28(8):507−15.

[7] Dominguez DD, Mowery RL, Turner NH. Friction and durability of well-ordered, close-packed carboxylic acid monolayers deposited on glass and steel surfaces by the Langmuir-Blodgett technique. Trib Trans STLE 1994;37(1):59−66.

[8] Levine O, Zisman WA. Physical properties of monolayers adsorbed at the solid−air interface-II. Mechanical durability of aliphatic polar compounds and effect of halo-genation. J Phys Chem 1957;61(9):1188−96.

[9] Bridgman PW. Recent work in the field of high pressures. Rev Mod Phys 1946; 18(1):1−93.

[10] Briscoe BJ, Scruton B, Willis FR. The shear strength of thin lubricant films. Proc Roy Soc Lond Ser A 1973;333:99−114.

[11] Campbell WE. Boundary lubrication. In: Ling FF, Klaus EE, Fein RS, editors. Boundary lubrication: an appraisal of world literature. New York, NY: ASME; 1969. p. 87−117.

[12] Rowe CN. Some aspects of the heat of adsorption in the function of a boundary lubricant. ASLE Trans 1966;9(1):101−11.

[13] Rowe CN. Role of additive adsorption in the mitigation of wear. ASLE Trans 1970;13(3):179−88.

[14] Stolarski TA. A system for wear prediction in lubricated sliding contacts. Lub Sci 1996;8(4):315−51.

[15] Memorandum on definitions and, symbols and units. Proc I Mech E 1957 London; 4.

[16] Glossary of terms and definitions in the field of friction, wear and lubrication. Paris: OECD, Research Group on Wear of Materials; 1969. p. 53.

[17] Ludema KC. A review of scuffing and running-in of lubricated surfaces, with asperities and oxides in perspective. Wear 1984;100:315−31.

[18] Blok H. Theoretical study of temperature rise at surfaces of actual contact under oil-iness lubricating conditions. Proceedings of the general discussion lubrication and lubricants, vol. 2. London: Institute of Mechanical Engineers; 1937. p. 222−35.

[19] Enrrichello R. Friction, lubrication and wear of gears. In: Blau P, editor. Friction lubrication and wear technology. Materials Park, OH: ASM International; 1992. p. 535−45.

[20] Matveevsky RM. The critical temperature of oils with point and line contact machines. J Fluids Eng ASME 1965;87:754−9.

[21] Leach EF, Kelley BW. Temperature—the key to lubricant capacity. ASLE Trans 1965;8(3):271−85.

[22] Bjerk RO. Oxygen—as an extreme pressure agent. ASLE Trans 1973;16(2):97−106.

[23] Ajayi OO, Hersberger JG, Zhang J, Yoon H, Fenske GR. Microstructural evolution during scuffing of hardened 4340 steel—implication for scuffing mechanism. Trib Int 2005;38(3):277−82.

[24] Ouyang Q, Okada K. A study of the quantitative description of the seizure behaviour of steels at low temperature in a vacuum. Trib Trans STLE 1998;41(2):301−5.

[25] Szlufarska I, Chandross M, Carpick RW. Recent advances in single-asperity nanotribology. J Phys D Appl Phys 2008;41(12):123001−39.

[26] Mate CM. Tribology on the small scale: a bottom up approach to friction, lubrication and wear. New York, NY: Oxford University Press; 1968 [chapter 10].

[27] Berman AD, Israelachvili JN. Microtribology and microrheology of molecularly thin liquid films. In: Bhushan B, editor. Modern tribology handbook, volume 1. Boca Raton, FL: CRC Press; 2001 [chapter 16].

[28] Leggett GJ. Friction force microscopy of self-assembled monolayers: probing molecular organisation at the nanometre scale. Anal Chim Acta 2003;479:17—38.

[29] Briscoe BJ, Evans DCB. The shear properties of Langmuir-Blodgett layers. Proc R Soc Lond Ser A 1982;380:389—407.

[30] Spikes HA. Direct observation of boundary layers. Langmuir 1996;12(19):4567—73.

[31] Fillot N, Berro H, Vergne P. From continuous to molecular scale in modelling elastohydrodynamic lubrication: nano scale surface slip effects on film thickness and friction. Trib Lett 2011;43(3):257—66.

[32] Chollet F, Liu HB. A (not so) short introduction to MEMS, version 5.1, 103, <http://memscyclopedia.org/introMEMS.html>; 2013.

[33] Gellman AJ, Spencer ND. Surface chemistry in tribology. In: Stachowiak GW, editor. Wear: materials, mechanisms and practice. Chichester: John Wiley; 2005. p. 110 [chapter 6].

[34] Mate CM. Atomic-force-microscope study of polymer lubricants on silicon surfaces. Phys Rev Lett 1992;68(22):3323—6.

[35] Williams JA, Le HR. Tribology and MEMS. J Phys D Appl Phys 2006;39(12): R201—14.

[36] Forbes ES. The load-carrying action of organo-sulphur compounds—a review. Wear 1970;15(2):87—96.

[37] Dorinson A, Ludema KC. Mechanics and chemistry in lubrication. Tribology Series 9. New York, NY: Elsevier; 1985.

[38] Sakurai T, Sato K. Study of corrosivity and correlation between chemical reactivity and load-carrying capacity of oils containing extreme pressure agents. ASLE Trans 1966;9(1):77—87.

[39] Godfrey D. Chemical changes in the steel surfaces during extreme pressure lubrication. ASLE Trans 1962;5(1):57—66.

[40] Tomaru M, Hironaka S, Sakurai T. Effect of oxygen on the load-carrying action of some additives. Wear 1977;41(1):117—40.

[41] Willermet PA, Kandah SK, Siegl WO, Chase RE. The influence of molecular oxygen on wear protection by surface-active compounds. ASLE Trans 1983;26(4): 523—31.

[42] Godfrey D. The lubrication mechanism of tricresyl phosphate on steel. ASLE Preprint, 64-LC-1; 1964.

[43] Klaus EE, Duda JL, Chao KK. A study of wear chemistry using a micro sample four-ball wear test. Trib Trans STLE 1991;34(3):426—32.

[44] Canter N. Special report: trends in extreme pressure additives. Tribol Lub Technol September, 2007:10—17.

[45] Nicholls MA, Do T, Norton PR, Kasrai M, Bancroft GM. Review of the lubrication of metallic surfaces by zinc dialkyl-dithiophosphates. Trib Int 2005;38(1): 15—39.

[46] Pereira G, Paniagua DM, Lachenwitzer A, Kasrai M, Norton PR, Capehart TW, et al. A variable temperature mechanical analysis of ZDDP-derived antiwear films formed on 52100 steel. Wear 2007;262(3—4):461—70.

[47] Minami I, Hasegawa T, Memita M, Hirao K. Investigation of antiwear additives for synthetic esters. Lub Eng STLE January, 2002:18—22.

[48] Morina A, Neville A, Priest M, Green JH. ZDDP and MoDTC interactions in boundary lubrication—the effect of temperature and ZDDP/MoDTC ratio. Trib Int 2006;39(12):1545—57.

[49] Fujita H, Spikes HA. The formation of zinc dithiophosphate antiwear films. Proc Inst Mech Engrs Part J J Eng Tribol 2004;218(4):265−78.

[50] Liu E, Kouame SD. An XPS study on the composition of zinc dialkyl dithiophosphate tribofilms and their effect on camshaft lobe wear. Trib Trans STLE 2014; 57(1):18−27.

[51] Bakunin VN, Suslov AY, Kuzmina GN, Parenago OP, Topchiev AV. Synthesis and application of inorganic nanoparticles as lubricant components—a review. J Nanoparticle Res 2004;6(2):273−84.

[52] Zhang L, Chen L, Wan H, Chen J, Zhou H. Synthesis and tribological properties of stearic acid-modified anatase (TiO$_2$) nanoparticles. Trib Lett 2011;41(2):409−16.

[53] Verma A, Jiang W, Safe HHA, Brown WD, Malshe AP. Tribological behaviour of deagglomerated active inorganic nanoparticles for advanced lubrication. Trib Trans STLE 2008;51(5):673−8.

[54] Martin JM, Ohmae N, editors. Nanolubricants. Tribology series. Chichester: John Wiley; 2008.

[55] Jaiswal V, Rastogi RB, Kumar R, Singh L, Mandal KD. Tribological studies of stearic acid-modified CaCu$_{2.9}$Zn$_{0.1}$Ti$_4$O$_{12}$ nanoparticles as effective zero SAPS antiwear lubricant additives in paraffin oil. J Mater Chem A R Soc Chem 2014;2(2):375−86.

[56] Rowe CN. Lubricated wear. In: Peterson MB, Winer WO, editors. Wear control handbook. New York, NY: ASME; 1980. p. 451−6.

CHAPTER 5

Fundamental Approaches to Chemical Wear Modeling

5.1 INTRODUCTION

Boundary lubrication without chemical additives can lead to catastrophic wear when the operating conditions become severe. This is because of increased metal contact through boundary films as discussed in the previous chapter. Antiwear additives can overcome this problem by reacting with the surfaces and developing wear resistant films. The control of wear by ZDDP in engines is an important example in this category. In some applications like neat drilling fluids, both EP and AW actions are needed from the additive package. The emphasis in this chapter is on chemical wear.

When reference is made to wear rate it has to be recognized that the wear can be very different during running-in and steady state. Since this issue will be taken up in detail in Chapter 7, it is not considered separately here. But it should be emphasized that literature values regarding wear rates in many cases are ill-defined since they are simply based on wear measured after a given period of running. Also several types of machines and operating conditions are involved with little scope for generalization.

First, consideration will be given to the new concepts being advanced in tribochemistry since they will have a bearing on wear modeling. The model based on parabolic rate law is then developed, followed by a careful assessment of the parameters involved. Based on this assessment it is proposed that the film removal can also be due to repetitive stressing, leading to fatigue wear. Implementation of the parabolic model is demonstrated by an example dealing with wear of cast iron. The next section will present other approaches to chemical wear modeling. This will include kinetic models based on surface reactions as well as a recent model that is based on stresses within the film. The final section summarizes the ideas related to chemical wear and emphasizes the importance of empirical modeling.

Modeling of Chemical Wear.
DOI: http://dx.doi.org/10.1016/B978-0-12-804533-6.00005-6
© 2016 Elsevier Inc.
All rights reserved. 105

5.2 ADVANCES IN TRIBOCHEMISTRY

Mechanochemistry may be defined as the chemistry involving mechanical energy as the input. When the energy input is due to frictional energy it is termed tribochemistry. Tribological contacts have two major effects on chemical reactions. One effect is due to the generation of nascent metal surfaces that are active. The other effect is due to the emission of exoelectrons in sliding and their interaction with surfaces and environment. The two effects are considered here separately. The treatment is limited to conceptual understanding. The relevance of these developments to the conventional (thermochemical) kinetic treatment of chemical wear is also discussed.

5.2.1 Surface Chemistry of Nascent Surfaces

During the wear process the oxide films tend to get removed and fresh metal surfaces interact with the environment. The environment consists of the lubricant/additives as well as dissolved oxygen. Besides this, pressure and contact temperature in the tribosystem also affect reactions. Even when oxides are removed they reform within a millisecond. Any research under atmospheric conditions cannot observe the effect of nascent surfaces exclusively. In fact earlier research in metal cutting where fresh metal surfaces are generated did not lead to definitive conclusions. To overcome this problem research is being conducted under ultra high vacuum (UHV) so that assured fresh metal surfaces can be generated and their reaction studied. The additives used are molecules that have required a functional group but can be easily vaporized and used in gaseous form. These tribological studies are coupled with XPS, AES, and mass spectrometry that provide clear information on the adsorption and reactions involved.

The UHV studies conducted in the past decade are revealing. While fresh metal surfaces are expected to be active it was found that activity of various compounds depended on the metal as well as the molecule. Some molecules show high activity while others do not. This puzzling behavior was explained qualitatively on the basis of Hard and Soft Acids and Bases (HSAB) principle. This concept, proposed by Pearson [1], was originally proposed to explain stability or otherwise of metal complexes. It has been extended to explain surface reactions as well. The acids and bases can be divided into hard and soft categories based on polarizability. Hard acids and bases are small and compact and nonpolarizable. Soft acids and bases are larger with more diffuse distribution of electrons. The classifications with regard to various organic groups and metallic surfaces are available in literature. The principle states that hard acids interact with hard bases while soft acids interact with soft bases preferentially. The principle can be elucidated by a simple example of inter-action between halogen and metallic ions. The halogen anions may be divided into hard, borderline, and soft bases as follows:

Hard F^- and Cl^-; Border line Br^-; Soft I^-

In a similar fashion the metallic cations can be classified in terms of acids. Four metallic ions are classified as:

Hard Li^+ and Na^+; Border line Fe^{2+} and Cu^{2+}; Soft Cu^+ and Ag^+

As per the Pearson principle it is easy to form LiF (hard−hard) and AgI (soft−soft) as compared to AgF and LiI. The organic functional groups also have been divided into hard, borderline, and soft categories. Their interaction with metallic surfaces can be explained on the basis of work reported by Mori [2], who conducted experiments in UHV. The nascent metal surface generated in UHV behaved very differently with regard to adsorption based on the HSAB principle. Considering the nascent steel surface as soft acid, soft bases adsorb strongly in comparison to hard bases. Hence propionic acid and propyl amine that are hard bases are weakly adsorbed while methyl propionate, which is a soft base, shows better adsorption. Among the EP additives the phosphate adsorption is poorer in comparison to disulfide. This again is due to the fact that phosphate is a hard base while the diethyl disulfide is a soft base. The author has also shown differences in adsorption between nickel and aluminum surfaces based on the HSAB principle by considering alumi-num and nickel as hard and soft acids. It may be noted that metal surfaces are classified into hard, borderline, and soft acids on the basis of their electronic configuration. This concept has also been recently invoked to explain the formation of borate glass in UHV utilizing gaseous trimethyl borate as reactant [3]. Oxide surfaces will behave differently to the nascent metals. For example, propionic acid, which is a strong base, adsorbs easily on iron oxide, which acts as a strong acid.

This research has not discussed the importance of triboemission on chemical reactions. When sliding takes place there is mechanical activa-tion leading to several processes. These include lattice defects, dangling bonds, emission of low energy exoelectrons, photons, and other particles. The exoelectrons, in $0-4$ keV range, have been found to be of particular importance in tribochemistry. When the electrons are emitted, positive sites are generated on the surface. The electrons in turn get attached to the lubricant/additive molecule leading to the formation of ions and free radicals. The negative ions interact with the positive sites of the surface, leading to chemisorbed layer. Kajdas has worked extensively in this area and his recent review [4] provides a detailed consideration of the reactions. A book on tribochemistry also deals with this and the HSAB principle [5]. A reference may also be made here to recent thorough work on tribochemical film formation by Philippon et al. [6]. The experiments were conducted in UHV with gaseous trimethyl phosphite. Iron phosphide was clearly detected in tribofilms though the temperature involved was room temperature. Phosphide formation by heating can only occur above 200 °C. The mechanism was explained on the basis of electron emission and is reproduced in Figure 5.1. In (a) surface oxide is removed due to

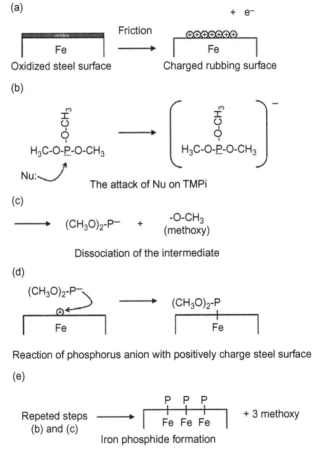

Figure 5.1 Mechanism of formation of iron phosphide on nascent metal when surface is reacted with trimethyl phosphite. *Reproduced from Ref. [6] with permission of D. Philippon. Copyright 2011 Elsevier Publishing.*

friction with generation of exoelectrons and corresponding positive sites on the surface. In (b) the electron interaction leads to formation of an intermediate that dissociates into methoxy and phosphide ion as shown in (c). The chemisorption of phosphide ion and further dissociation to phosphide are shown in (d) and (e).

The interaction of anion and cation is a Lewis acid—base interaction. Note that activation energies involved in this case will be far lower than the thermal route. This discussion is based on UHV conditions. In normal atmosphere the relative importance of HSAB and exoelectron interaction will depend on the operating conditions, material combination, and environment. It is difficult in such cases to assign the mechanism. Sometimes these possibilities are invoked to explain unexpected behavior. As pointed out in the previous chapter film formation with ZDDP at room temperature was considered possibly due to catalytic effect of nascent iron surface and/or reactions induced by exoelectrons [7]. The issue of importance with regard to the present book is the relevance of these mechanisms to kinetic treatment of chemical wear. This issue will be considered when available kinetic models are discussed.

5.3 KINETIC MODELING OF CHEMICAL WEAR BASED ON CRITICAL THICKNESS

5.3.1 The Basic Model

The chemical wear modeling is based on the film wear concept of Archard [8]. In this model wear is assumed to occur when a critical film thickness is reached. The film growth rate is obtained by kinetic modeling of the reaction. By linking critical thickness to the growth rate, the final wear rate equation is obtained. The steps in derivation are explained below.

First consider adhesive wear equation, Eqn (3.23), defined by

$$\frac{V}{l} = K\frac{W}{H}$$

where K is the wear coefficient. This equation is derived on the basis of the assumption that a hemispherical wear particle is removed from asperity contact with diameter d over a sliding distance d. Each contact does not result in a wear particle. In fact the probability of removal is low and this is accommodated by the nondimensional wear coefficient K.

In the case of film wear it is considered that wear occurs at an asperity when a critical film thickness ξ is reached. The wear volume of the particle will be equal to the product of asperity contact area and thickness. The volume generated over sliding distance d at one asperity is then

$$v_d = \frac{\pi d^2}{4} \xi \qquad (5.1)$$

In this equation note that this volume of wear occurs *after* a critical thickness ξ is reached over a given number of contact cycles.

Following arguments similar to derivation of adhesive wear equation it can be shown

$$\text{Wear rate } V_r = \frac{V}{l} = K_f \frac{\xi}{d} \frac{W}{H} \qquad (5.2)$$

In this equation K_f is the wear coefficient that is equal to the inverse of number of cycles needed to build a film thickness ξ. K_f can be derived based on kinetic considerations. This is unlike K for adhesive wear, which can be found only experimentally.

Assume the total time needed to build critical thickness is t. The time during one encounter t_c is d/v, where v is the sliding velocity. K_f, the inverse of total number of encounters, will be

$$K_f = \frac{t_c}{t} = \frac{d}{vt} \qquad (5.3)$$

Static reactions with chemical additives as well as oxidation follow parabolic law. As per this law the increase in film thickness expressed in terms of mass change per unit area Δm, expressed in kg/m^2, will be related to the reaction time t as

$$\Delta m^2 = k_p t \qquad (5.4)$$

where k_p is the rate constant (kg^2/m^4/s).

From kinetic considerations it is known

$$k_p = A_p \exp\left(\frac{-Q_p}{RT_c}\right) \qquad (5.5)$$

where
 A_p = Arrhenius constant (kg^2/m^4/s),
 Q_p = activation energy (J/mol),
 T_c = contact temperature (K),
 R = gas constant (J/mol-K).

Expressing mass change in terms of density ρ (kg/m^3) and taking f as the fraction of the volume that is reacted product it can be shown

$$K_f = \frac{dA_p \exp\left(\frac{-Q_p}{RT_c}\right)}{\xi^2 \rho^2 f^2 v} \tag{5.6}$$

Substituting for K_f in Eqn (5.2)

$$\text{Wear rate } V_r = \frac{V}{l} = \frac{A_p \exp\left(\frac{-Q_p}{RT_c}\right)}{\xi \rho^2 f^2 v} \frac{W}{H} \tag{5.7}$$

Usually complete reaction is assumed and so f is taken as unity. With complex additives the reaction films cannot be easily characterized and kinetics may be based only on film thickness as discussed later in Section 5.4.2.

One specific application of this approach is to oxidational wear. Quinn [9–11] published several papers on oxidative wear. His approach is based on adhesive wear equation of Archard. His derivation of K is equivalent to the derivation of K_f as described earlier. He then used the adhesive wear equation of Archard with this derived K. His equation then comes out as

$$V_r = \frac{V}{l} = K \frac{W}{H} = \frac{dA_p \exp\left(\frac{-Q_p}{RT_c}\right)}{\xi^2 \rho^2 f^2 v} \frac{W}{H} \tag{5.8}$$

Equation (5.8) can be obtained by multiplying Eqn (5.7) with d/ξ. It is to be noted that Archard's adhesive wear equation is based on the removal of a hemispherical particle of volume $(1/12)\pi d^3$ over a distance d. When chemical reaction is involved the wear volume will be based on the film removal. So the volume removed at an asperity will be $(\pi d^2/4)\xi$. The present authors consider that their approach is justified and hence will use this approach for chemical wear. Note that as per our equation wear rate is independent of d.

Asperity contacts keep varying and the film build-up varies. For example, in a pin-on-disc machine the film builds faster on the stationary pin in comparison to the rotating track on which the film is distributed over a larger area. But after the wear process continues over a fairly long period the wear rate is governed by Eqn (5.7) for both surfaces.

5.3.2 Discussion of the Critical Thickness Model

The nature of equation and the parameters need careful consideration. First the reaction parameters involved are based on static reaction at different temperatures. For static reaction Q_p the activation energy, and A_p the Arrhenius constant, are considered to be constant. Also the reaction temperature is easily controlled and measured.

In a dynamic wear situation the activation energies involved for reaction can be different and may not be constant. The contact temperature T_c involved in the reaction at the asperities is difficult to predict. Also reaction may be influenced by the temperature surrounding the asperities. The concept of critical film thickness also needs elaboration. These issues are considered next.

5.3.2.1 Activation Energy

It is first necessary to take into account the modern developments in tribochemistry discussed in Section 5.2. Some chemical reactions with nascent surfaces can occur very easily because of exoelectrons as exemplified in Figure 5.1. Such reactions will have low activation energies in comparison to thermally activated reactions. To exemplify the influence of activation energy consider Eqn (5.7) for wear rate. Taking an example of oxidational wear [10] the value for static reaction is 193 kJ/mol. For tribooxidation in the temperature range of 100−300 °C the value of activation energy estimated is ∼26 kJ/mol. This makes a huge difference to wear rate. The ratio of wear rates at low and high activation energies can now be calculated taking R to be 8.31 J/K/mol assuming Arrhenius constant is same in both cases as follows:

$$\exp(-26,000/8.31 \times 373)/\exp(-193,000/8.31 \times 373) = 2.50 \times 10^{23}$$

This value is very high but exemplifies the very large influence of activation energy. However the values of Arrhenius constants can be very different for tribological contacts and static contacts. In tribooxidation both activation energy and Arrhenius constant vary with operating conditions and nature of oxides. These problems lead to a lot of models for oxidation, none of which are capable of wear prediction from first principles. Leaving aside actual values the important point that comes out is that tribocontacts can accelerate the reaction substantially in comparison to static reaction. The wear rates will become very high unless the reaction films form a strong barrier for further reaction. The issue is further examined when the aspect of critical film thickness is considered. The initial reaction can in some cases happen with low activation energy. The subsequent reactions leading to film build-up can be a combination of thermal and surface activation.

Now assuming that a particular activation energy is effective in the dynamic reaction the question is how to obtain the value. Known procedures are available to obtain the energy of activation provided the basic wear equation is applicable. This will be illustrated by analyzing the experimental work done by the authors in Section 5.3.2.4.

5.3.2.2 The Contact Temperature T_c

The temperature rise due to friction when added to the bulk temperature is the contact temperature. In the lubricated contact the bulk temperature is taken as the lubricant temperature. As discussed in Section 3.3.2 the temperatures may be specified as follows.

Bulk temperature T_b = oil temperature,

Surface temperature $T_s = T_b$ + temperature rise of the geometric area $\Delta\theta_g$,

Asperity contact temperature $T_c = T_s$ + temperature rise at the asperity $\Delta\theta_a$.

The temperature rise can be related to either maximum or average or both. The problem of temperature effect can be illustrated as shown in Figure 5.2.

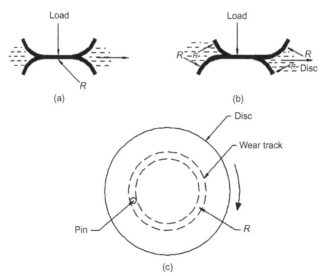

Figure 5.2 Chemical reaction zones represented by R. (a) Reaction at high asperity temperature; (b) reaction in the asperity cooling zone; and (c) bulk reaction on the wear track.

In Figure 5.2 the situation of contact and separation are shown. In (a) the contact over a single asperity is shown while in (b) the asperity is going out of contact. In (c) the overall contact track is shown. The chemical reaction can occur at the elevated asperity temperature as well as surface temperature. While asperity temperature is high the duration and the availability of the additive in the contact zone are the limiting factors. The asperity will have adsorbed additive over the reacted film as it enters into contact. These layers with the superimposed bulk lubricant enter the contact where the lubricant is squeezed out leaving probably monolayers of the reactant. In such a case the reaction may be limited to available monolayers despite high temperature. But as the asperity comes out of contact and starts cooling, continued reaction is possible due to easier availability of additive through diffusion. The geometric patch soon enough reaches the bulk temperature. Reaction can occur at bulk temperature as well provided the temperature is high enough. In view of this a careful assessment of temperature problem is necessary. The problem is difficult and often the temperature at asperity is assumed to govern the reaction.

5.3.2.3 The Concept of Critical Thickness

Detailed research has been done over a period of two decades by Quinn and his associates, as mentioned previously. With regard to several chemical additives used, only few kinetic treatments are available. While extensive work is being done on the chemistry of ZDDP films the work related to wear is limited to a comparison of the influence of different types of ZDDP and synergism or antisynergism effects with different additives. Taking the example of oxidational wear at high temperatures (>300 °C) it was observed that elevated pads form during the wear process with thickness in the range of $1-2\,\mu m$ [10,11]. These films seem to detach due to fatigue cracks. With reference to ZDDP films observations have shown that elevated pads with $80-100$ nm thickness are involved in lubrication as shown in Figure 4.10. No clear information is available regarding how these films detach causing wear. We can only guess two possible routes to film removal. In one case the film is removed as soon as a critical thickness is reached. The second possibility is that a critical thickness really amounts to a situation where further growth of the film is negligible. After this the film undergoes cyclic stressing and finally gets detached. The second possibility seems feasible at least with ZDDP films that can have nanohardness as high as 3 GPa. The tenacity of the ZDDP film is also evident from the work of Fujita and Spikes [7], who formed tribofilm first in the presence of an additive and then ran a further test with base oil alone. It is remarkable that tribofilm did not wear out for several hours. It is hence possible that the wear can be due to fatigue and needs a large number of cycles before detachment.

The possibility of fatigue wear can also be supported by analyzing the parabolic growth law with a reasonable assumed wear rate. Wear of additives has been expressed in terms of K as in Archard's wear equation for comparison purposes. The K values typically range from 0.1 to 10×10^{-8} as per earlier literature [12]. As ZDDP is a very good antiwear additive it is reasonable to take K as 10^{-8}. This value is based on measurement of the wear loss of material that is steel. If it is assumed that metal content in worn film is half the volume then film wear rate should be twice this value. Hence wear coefficient for film expressed as K_1 will be $2K$ and is equal to 2×10^{-8}. It is necessary to express this value in terms of K_f for film wear. As per Eqn (5.2) $K_1 = K_f(\xi/d)$. Taking the critical thickness as equal to typical pad thickness of 100 nm and pad diameter of 10 μm the value of K_f will be 2×10^{-6}. This amounts to a growth of 0.1 μm in 0.5×10^6 cycles. Now let the parabolic growth law be recast as follows:

$$\Delta r^2 = k_n n \tag{5.9}$$

where n is the number of cycles, Δr is film thickness in μm, and k_n is a dimensional constant.

The value of Δr will be 0.1 μm for $n = 1/K_f = 0.5 \times 10^6$ cycles. Hence the value of k_n will be 2×10^{-8}.

Differentiating Eqn (5.9)

$$2\Delta r \frac{\mathrm{d}\Delta r}{\mathrm{d}n} = k_n \tag{5.10}$$

where $\mathrm{d}\Delta r/\mathrm{d}n$ is the film thickness growth per cycle.

So at 0.5×10^6 cycles the growth rate is as low as 10^{-7} μm per cycle. This is a very low value of 0.001 Å per cycle! If the growth rate for the same reaction is evaluated at 10^3 cycles the value of per cycle growth rate will be 2.24×10^{-6} μm and the growth rate is as low as 0.0224 Å. While the rates will vary depending on selected values of d and ξ (within reasonable limits) the changes still mean very low unrealistic growth rates. This situation can only be reconciled with quick growth of film followed by a strong barrier for further reaction. The wear then occurs over a large number of fatigue cycles resulting in wear. Hence film wear by fatigue is likely to be an important component of wear process. The number of fatigue cycles will depend on the nature of film that forms in the initial cycles and should be governed by the normal and shear stresses in the film. Another important issue is where the film failure occurs in the film. Referring to the typical film shown in Figure 4.10 the removal may occur in the upper zone or the lower zone of the film. If the removal is confined to the upper part of the film then wear will be very low since removal involves mainly sacrificial film with low iron content. On the other hand removal within the short-chain zone will result in high wear as more iron will be removed with the film. Further assessment on these lines is not possible due to lack of information. But this argument brings out two aspects that need to be investigated for developing wear models.

The first step is to see how tribofilm behaves when subjected to static reaction at the micro regions. If the reaction rate is negligible it supports the concept of barrier film. The other issue is the characterization of the film with regard to fatigue. It may be possible to assess this property by cyclic stressing of the film as per the procedure adopted by Karmakar [13,14]. Other experimental methods for assessment of fatigue may also be possible. Such characterization of film response to dynamic stressing may give a much better insight into film wear. It may be pointed out here that a new approach to wear based on mechanical properties of the film and substrate has been proposed [15] and shall be discussed in Section 5.4.2. Note that all these remarks are centered around ZDDP, on which several investigations are available. Several other antiwear additives used like phosphorous compounds and sulfur compounds may also involve similar behavior in view of the low wear rates involved. But less information is available on these additives and further assessment is not possible. The schematic shown in Figure 5.3 explains the proposed wear mechanism for ZDDP additives.

With highly reactive sulfur additives used for EP action this mechanism may not be applicable. The thick reaction films may easily crack and wear away, which is typical for corrosive films. This reaction is not allowed to continue in practice because the EP additive acts only for a short duration when scuffing conditions arise. Similarly mild and severe oxidational wear films that form at high temperatures can also be easily removed when a critical thickness is reached.

5.3.2.4 Example of Implementation of Critical Thickness Model

The kinetic model is applied to the wear of gray cast iron and is partly based on the earlier work reported by the authors [16]. Activation energy Q_p and Arrhenius constant A_p have been estimated and their relevance

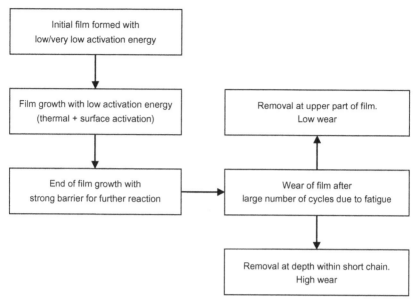

Figure 5.3 Proposed wear mechanism for ZDDP additive. A similar mechanism may be involved for other antiwear additives.

discussed. Experiments were conducted in a reciprocating rig with a piston ring cut piece moving against a stationary liner piece. Both specimens were made of cast iron. The tests were conducted with load W and bulk temperature T_b as variables. Three different loads of 30, 50, and 70 N and bulk temperatures of 50, 100, and 150 °C were used. A total of 11 experiments based on experimental design were conducted, in which two repeat experiments were conducted at mean level of load and temperature. The frequency was 50 Hz and stroke length was 1 mm. This amounts to a linear velocity v of 0.1 m/s. The wear was recorded as a weight loss at 5, 10, 15, 20, and 25 min. It was converted into wear volume by taking density of cast iron as 7150 kg/m^3. SEM examination showed the removal involved thin flaky particles that were assumed to be oxides. Running-in and steady state wear rates at each operating condition have been characterized by using the program developed by the authors [17]. The details of the program are given in Chapter 7. The running-in wear is a combination of adhesive and oxidative wear, hence it cannot be analyzed on the basis of the kinetic equation. Hence the steady state wear of ring and liner where oxidational wear is assumed has been modeled here. Equation (5.7) for wear modeling can be recast as follows taking logs on both sides. It may be noted that v here has a constant value and is accommodated into the constant a_0. The constant a_1 is used to accommodate any possible exponent on W.

$$\log_e(\text{Sw}) = a_0 + a_1 \log_e(W) + \frac{a_2}{T_c} \qquad (5.11)$$

This equation is now treated as an empirical equation and the various parameters are estimated. The coefficients a_0, a_1, and a_2 were found by the least square method. The detailed methodology is given in Chapter 6. It should be noted that temperature T_c at different operating conditions was found by adding the temperature rise $\Delta\theta$ based on geometric area to the bulk temperature T_b. The value of $\Delta\theta$ ranged from 4 to 6 °C. The empirical relationships found for ring and liner are given below.

Steady state wear rate in mm^3/mm for ring

$$\text{Sw}_{\text{ring}} = 9.97 \times 10^{-7} W^{0.88} \exp\left(-\frac{642.44}{T_c}\right) \qquad (5.12)$$

Steady state wear rate in mm^3/mm for liner

$$\text{Sw}_{\text{liner}} = 2.09 \times 10^{-8} W^{1.52} \exp\left(-\frac{725.57}{T_c}\right) \qquad (5.13)$$

Correlation coefficients for the ring and liner were found to be 91% and 89%, respectively.

Comparing these equations with Eqn (5.7), activation energy Q_p for the ring and liner were found to be 5.34 and 6.03 kJ/mol, respectively, by equating Q_p/R to 642.44 and 725.57. The activation energy estimate from the above systematic statistical procedure and comparison with an analytical kinetic equation for ring and liner are close at steady state, hence it may be concluded that the main type of wear is oxidational. Dynamic activation energy of this combination of cast iron surfaces may be taken as an average of these two values that is equal to 5.69 kJ/mol. The above treatment shows that the kinetic model based on parabolic oxidation is reasonably valid in the present case. The activation energy found is much lower than static oxidation and this has been observed by Quinn, also as discussed in Section 5.3.2.1. For example, the activation energy for steel surfaces under static conditions can range from 96 to 210 kJ/mol depending on the nature of oxides formed and temperature range [18]. This is also true for chemical additives and is discussed in Section 5.4.1. The main reason for low activation energy is the tribo-activation of the surfaces.

Arrhenius constant A_p estimation is demonstrated by considering the steady state wear rate of liner at 50 N and 378 K using activation energy for a liner of 6.03 kJ/mol. The wear rate under these conditions calculated from Eqn (5.13) was 1.17×10^{-6} mm^3/mm. Then the wear coefficient K of liner was found by using the Archard equation, Eqn (3.23). The value was found to be 5.67×10^{-5} by taking the hardness of cast iron as 2.42 GPa. The next step is to find K_f by multiplying (d/ξ) with K. If the asperity contact diameter d is taken as 10 μm and typical critical film thickness ξ as 0.5 μm the value of K_f is 1.13×10^{-3}. Now using Eqn (5.6) the Arrhenius constant A_p may be obtained. Taking f as unity, which amounts to an assumption that worn material is fully oxidized, the average density of oxide ρ as 5240 kg/m^3, gas constant R as 8.31 J/mol-K, the A_p was calculated as 5.30×10^{-4}. If ξ is 1 μm the A_p will be 1.06×10^{-3}. Similarly A_p was calculated at different operating conditions; it ranges from 4.07×10^{-4} to 6.32×10^{-4} for ξ of 0.5 μm and it ranges from 8.14×10^{-4} to 1.26×10^{-3} for ξ of 1 μm. Equation (5.7) shows that the Arrhenius constant is proportional to critical film thickness and is independent of the asperity contact diameter. The critical film thickness is difficult to quantify exactly. If it can be clearly quantified then A_p can be estimated from the above procedure.

The significance of A_p values obtained from a fundamental point of view is difficult to ascertain. Unlike chemical reactions in which the A_p value is nearly constant over a wide range of operating conditions the value obtained here is very dependent on operating conditions. It is indirectly derived based on K_f, ξ, d, and v. It may be best to model wear on the basis of Eqns (5.12) and (5.13) with the derived overall constant. However it may be noted that activation energy has a fairly constant value over a wide range of temperatures and is consistent with the theory.

The purpose here is mainly to show how the critical thickness model is implemented based on parabolic law. If the experimental data do not fit this, other laws like logarithmic growth law or purely empirical approaches may need to be tried.

5.4 OTHER APPROACHES TO WEAR MODELING

5.4.1 Kinetic Model Based on Surface Reaction

The early effort in kinetic treatment of chemical wear was due to Okabe and others [19] using radioactive techniques to monitor sulfur content on the surface. In these studies wear due to sulfur compounds was studied by expressing formation as well as removal of sulfides by kinetic rate equations. The following equations for formation and removal for elemental sulfur solution show the approach. It should be stated that the model is based on chemisorption and is not consistent with the reality of thick reaction films that form in such contacts.

Formation

$$Fe + S \rightarrow FeS$$

Formation rate of FeS is $v_1 = k_1[Fe][S]$ where k_1 is the rate constant with iron and sulfur concentrations expressed in mol/cm^2. In all the equations that follow concentrations are expressed in mol/cm^2.

The rate of removal of FeS is expressed as

$$v_2 = k_2[FeS]$$

where [FeS] is the surface concentration of iron sulfide.

The net formation rate of FeS may be now expressed as

$$\frac{d[FeS]}{dt} = k_1[Fe][S] - k_2[FeS] \tag{5.14}$$

If m is the original iron content then iron content at any instant is

$$[Fe] = m - [FeS] \tag{5.15}$$

The sulfur concentration [S] is taken to be a constant value l assuming adsorption is in equilibrium with the sulfur solution. Taking [FeS] as C, Eqn (5.14) may be written as

$$\frac{dC}{dt} = k_1 l(m - C) - k_2 C \tag{5.16}$$

when steady state is attained at $t = t_\infty$ net formation rate of FeS is zero and so $dC/dt = 0$. Integration of Eqn (5.16) can be done and the constant of integration can be obtained at the boundary condition $t = \infty$. Further simplification is possible taking $dC/dt = 0$ when $t = \infty$ and $C = C_\infty$. Here C_∞ refers to steady state concentration and is a constant. This leads to the following equations:

$$C = C_\infty \left(1 - e^{-(k_1 l + k_2)t}\right) \tag{5.17}$$

$$C_\infty = \frac{k_1 l m}{k_1 l + k_2} \tag{5.18}$$

These equations can be used to find kinetic parameters based on experiments. Further the cumulative formation of FeS per unit area in time t, expressed as W/A, has been obtained by integrating the formation rate equation and substituting for C from Eqn (5.17). The equation is

$$\frac{W}{A} = k_2 C_\infty t + \frac{k_1 l C_\infty}{k_1 l + k_2} \left(1 - e^{-(k_1 l + k_2)t} \right) \tag{5.19}$$

It can be shown that steady state wear rate q_s at $t = \infty$ is equal to $k_2 C_\infty$, while initial wear rate q_i at $t = 0$ is equal to $k_1 lm$. Equation (5.19) can now be written as

$$\frac{W}{A} = \frac{q_i - q_s}{k_1 l + k_2} \left(1 - e^{-(k_1 l + k_2)t} \right) + q_s t \tag{5.20}$$

As observed by Kloss and Wasche [20] this type of equation is very similar to the empirical equation used by the present authors for cumulative wear volume $V(t)$ as given below. The steady state and initial wear rates will now have unit of volume/s

$$V(t) = \frac{q_i - q_s}{b} (1 - e^{-bt}) + q_s t \tag{5.21}$$

where b is a constant.

The similarity is obvious if $(k_1 l + k_2)$ is replaced by the constant b.

The empirical equation is based on a realistic assumption that the running-in wear will be higher and does involve some metallic wear. The wear rate reduces with time as film formation takes over. In this empirical model also steady state is assumed at $t = \infty$. A detailed consideration of the empirical equation is taken up in Chapter 7.

The kinetic treatment above is simply dealing with wear in terms of FeS only. In many cases it is difficult to know the kinetics of reaction when complex molecules are involved. A similar kinetic approach has also been used to study dry wear, taking into account third-body formation [21]. These approaches do not specify how the material is removed. It is only based on a mass balance. As considered earlier, triboactivation definitely leads to faster reaction as compared to thermal activation. A formal development of this area is being attempted and is inspired by the original detailed consideration of mechanochemistry by Heinicke [22]. Significant work is being reported by Russian researchers and is based on active collision theory and activated complex theory [23]. However the present authors are not in a position to evaluate the utility of these ideas in terms of wear modeling.

More recently Fujita and Spikes [7,24] have applied kinetics to study the formation and removal of tribofilms formed in a mini traction machine. These tests were conducted with primary and secondary ZDDP. With secondary ZDDP film forms quickly, reaches a maximum, and then decreases. The film build-up could be modeled on the basis of first-order kinetics while decrease in film thickness attributed to wear was modeled empirically. The basic concept involved in kinetic treatment by the authors may be illustrated as follows.

Taking the simplest model the film formation rate may be expressed by first-order kinetics as follows:

$$\frac{dX}{dt} = (1 - X)k_1 \qquad (5.22)$$

where

X = fraction of rubbed surface covered by film,

k_1 = rate constant.

Further assuming an inhibition time t_i with $X = 0$ it can be shown

$$X = 1 - e^{-k_1(t - t_i)} \qquad (5.23)$$

Assuming an asymptotic maximum thickness of h_{max}, average film thickness at a given instant is given by

$$h_{mean} = h_{max}(1 - e^{-k_1(t - t_i)}) \qquad (5.24)$$

where h_{max} is the experimentally obtained value.

The film removal rate in dispersant oil follows equation of the form

$$\frac{dX}{dt} = -k_5(X - X_0)^n \qquad (5.25)$$

where k_5 is a rate constant, X_0 is an irremovable residual fraction of rubbed surface covered by film, and n is an exponent.

The fit with experimental values could be obtained with $n = 4$. The rates were then expressed in terms of average film thickness and used to model the experimental behavior of the secondary ZDDP. Treatment of data by this type of relationship can also be considered empirical fits with appropriate constants.

The primary ZDDP forms film slowly and reaches steady value in 4 h. The tribofilms obtained with primary ZDDP were subjected to a wear test with base oil alone. Very little wear occurs over a period of hours, showing films are very tenacious and difficult to remove.

Estimation of activation energy gave a value of 4 kJ/mol/K, which clearly shows triboactivation. Also it is of interest to note that temperature dependence of the rate constant is linear unlike the expected exponential relation for thermally activated reactions. The formation and removal rates refer to the films. Wear of the steel specimens should be governed by the iron content of the removed films.

The actual removal mechanism remains unknown. Kinetic factors for formation and wear can be obtained without knowing the removal mechanism. The removal process of the film will strongly affect the wear rate of the material and should be understood. An interesting recent paper that dealt with wear of cam lobes in a real system [25] is indicative in this connection. The authors measured wear by profilometry at different positions of the lobe over a long duration of operation. The commercial formulation contained secondary ZDDP. Examination of wear scars by XPS showed that high wear zones were covered with short-chain phosphates while low wear zones were covered with long-chain phosphates. The authors interpreted this to mean that long-chain phosphates give reduced wear while short-chain phosphates give higher wear. An alternative explanation can be considered. As short chains are involved closer to the metallic surface, higher wear may be due to removal of the film at a depth closer to the metallic surface with higher iron content. If, on the other hand, the film wears preferentially within the upper long-chain zone lower wear is to be expected. This possible explanation fits into the earlier conceptual model given in Figure 5.3.

5.4.2 Stress-Based Wear Modeling of Reaction Film

The mild wear in boundary contact has been recently modeled by Bosman and Schipper [15]. The mild wear situation is defined as a case in which the wear occurs due to film removal in each asperity contact, which is later replenished by a reaction with an additive. An elaborate elasto-plastic contact model has been used to estimate the plastic strain in the reaction film. As the reaction layer is <100 nm the plane stress condition is assumed; that is, there is no stress variation in the layer along a normal direction. Normal stresses in the reaction layer are calculated taking elastic properties of the layer and considering the strain at the boundary of bulk material in which no slip condition between layer and bulk material is assumed. The strain at the boundary of bulk material is estimated by taking bulk material properties. Shear stress is estimated using a typical friction coefficient. The elastic properties of the film needed for the modeling have been obtained from literature in which nanoindentation studies were reported. The detailed model is not discussed here.

In each contact, material equivalent to a penetration depth δ is assumed to be removed. The penetration depth is related to plastic strain in thickness direction as follows:

$$\delta = \varepsilon_{zz}^{\text{layer}} h_{\text{balance}}$$

where h_{balance} is the steady state film thickness assumed to be constant. The wear volume at any discrete element will then be $\delta \Delta x \Delta y$. This volume is the film volume and it is necessary to know the volume of iron in the film. The authors used the XPS depth profiling available in literature and developed an empirical equation to obtain a volumetric percentage of iron as a function of depth $W_{\text{perc}}(\delta)$. Hence the wear volume in terms of iron loss will be $W_{\text{perc}}(\delta)\delta \Delta x \Delta y$.

This model is a good approach but appears to be in conflict with one experimental observation. As stated in Section 5.4.1, Fujita and Spikes [7] subjected tribofilms formed with primary ZDDP to further running with base oil alone without additives. They found that film did not wear out for several hours. As per the above theory wear should occur at each contact and because there is no additive in base oil there will be no reformation of the film. Such a situation should result in fast removal of film. It may be argued that the films formed are tenacious and get removed only after large number of stress cycles by fatigue. This is consistent with the model described in Figure 5.3.

There is a clear need to understand the fatigue behavior of ZDDP tribofilms. This can be approached by cyclic stressing of the tribofilm with different stress levels. One example of such work is due to Karmakar et al. [13,14] as mentioned earlier. They subjected medium carbon steel surfaces to different levels of cyclic stressing in the presence of lubricant. The stressed surface was then subjected to one pass sliding. From the wear behavior it was inferred that major crack propagation occurs only after a particular number of cycles are reached. The behavior was related to the surface tensile stresses. Direct evidence for crack propagation was provided by Ravi and Sethuramiah [26] by observing acoustic emission as a function of the number of stress cycles. The acoustic emission increased to a high level once a threshold level was crossed. This work was with bulk material and not film. The idea can be adopted to observe the behavior of films but it needs research to develop a suitable methodology.

The data can be related to experimental wear behavior and can form an important link between the nature of films and their wear behavior.

5.5 SUMMARY OF THEORETICAL MODELS AND NEED FOR EMPIRICAL APPROACH

Kinetic approaches of two types have been described. One approach is based on the growth of thickness to a critical level followed by removal. Another approach considers kinetics on the basis of surface reaction only. The first approach referred to as a critical thickness model is more realistic as invariably films of certain thickness are involved. This wear model has been extensively studied with regard to high temperature oxidation with limited success. With regard to chemical additives the major research has been with regard to characterization of ZDDP tribofilms. But wear modeling on the basis of critical thickness has not been attempted.

The major issues involved in modeling are:

1. There is a large difference in the activation energy of static and dynamic reactions. The main reason for this is trioactivation of surfaces as discussed in Section 5.2.1. The activation energies estimated can be as low as 10−20% of the static value. This means that reactions occur much faster in dynamic situations. The wear will be unrealistically large if film removal is assumed as soon as a critical thickness is formed. Dry oxidational wear is the only case where this theoretical model has limited success.

2. The early models simply assumed that film gets removed when a critical thickness is achieved without a detailed consideration of the actual mechanism. As discussed in Section 5.4.1 antiwear additives form tenacious films very quickly. But these films are removed after a large number of cycles. The most likely mechanism for removal is fatigue wear. In fact fatigue wear may be the rate determining step.

3. While fatigue modeling is important the work reported in this area as discussed in Section 5.4.2 is limited and cannot be effectively applied. But further developments may provide viable models that can be used.

These considerations show that theoretical modeling is not possible at this stage. On the other hand there is an urgent need to distinguish additives with regard to wear in view of the technological need for replacement of conventional additives. There is also a need for systematic optimization of additive packages. The later three chapters shall consider the detailed approaches as developed by the authors.

NOMENCLATURE

a_0, a_1, a_2 constant
A_p Arrhenius constant
b constant
C surface concentration, mol/cm^2 of iron sulfide
C_∞ steady state surface concentration, mol/cm^2 of iron sulfide
d average asperity contact diameter
f fraction of volume that is reacted product
[Fe], [S]   surface concentration, mol/cm^2 of iron and sulfur, respectively
[FeS]   surface concentration, mol/cm^2 of iron sulfide
$h_{balance}$ steady state film thickness of reaction film
h_{max} asymptotic maximum film thickness
h_{mean} average film thickness
H hardness of softer material
k_n rate constant in parabolic growth law adapted to number of cycles
k_p rate constant for parabolic growth
k_1 rate constant for formation of iron sulfide
k_1 rate constant for formation of ZDDP film
k_2 rate constant for removal of iron sulfide
k_5 rate constant for removal of ZDDP film
K adhesive wear coefficient, nondimensional number
K wear coefficient, nondimensional number, based on measurement of the wear loss of material that is steel
K_1 wear coefficient, nondimensional number, based on reacted film that contains material loss as steel along with reacted product, the value is greater than K
K_f film wear coefficient equal to inverse of cycles needed to form critical film thickness ξ
l constant surface concentration, mol/cm^2 of sulfur
l sliding distance
m original iron content per unit area
n number of cycles
n exponent
q_i initial wear rate at $t = 0$
q_s steady state wear rate at $t = \infty$
Q_p activation energy
R gas constant
R chemical reaction zone as shown in Figure 5.2
Sw steady state wear rate
Sw_{ring} steady state wear rate for ring
Sw_{liner} steady state wear rate for liner
t reaction time
t time needed to form film of critical thickness ξ
t_c time during one encounter of asperities
t_i inhibition time
t_∞ time to reach steady state surface concentration of iron sulfide
T_b bulk temperature

T_c	asperity contact temperature
T_c	contact temperature
T_s	surface temperature
v	sliding velocity
v_d	wear volume generated over sliding distance d at one asperity
v_1	formation rate of FeS
v_2	removal rate of FeS
V	wear volume
$V(t)$	cumulative wear volume in time t
V_r	wear rate, wear volume/sliding distance
W	load
$W_{perc}(\delta)$	volumetric percentage of iron as a function of depth δ
$\frac{W}{A}$	cumulative formation of FeS per unit area in time t
X	fraction of rubbed surface covered by ZDDP film
X_0	irremovable residual fraction of rubbed surface covered by ZDDP film

Greek Letters

δ	plastic penetration depth of the reaction film in contact
$\delta \Delta x \Delta y$	wear volume at any discrete element of area $\Delta x \Delta y$
Δm	mass change per unit area
Δr	film thickness
$\Delta \theta$	temperature rise based on geometric contact area
$\Delta \theta_a$	asperity temperature rise in contact
$\Delta \theta_g$	temperature rise in the geometric contact area
ε_{zz}^{layer}	plastic strain in thickness direction
ρ	average density of reacted film
ξ	critical film thickness

REFERENCES

[1] Pearson RG. Chemical hardness: applications from molecules to solids. Weinheim: Wiley-VCH; 1997.

[2] Mori S. Boundary lubrication from the viewpoint of surface chemistry-role of nascent surface on tribochemical reaction of lubricant additives. JTEKT Eng J English Edition 2011;1008E:1−12.

[3] Philippon D, De Barros-Bouchet MI, Lerasle O, Le Mogne T, Martin JM. Experimental simulation of tribochemical reactions between borates esters and steel surface. Tribol Lett 2011;41(1):73−82.

[4] Kajdas C. General approach to mechanochemistry and its relation to tribochemistry. In: Pihtili H. editor. Tribology in engineering; 2013. ISBN 978-953-51-1126-9, [chapter 11].

[5] Pawlak Z. Tribochemistry of lubricating oils. Tribology Series 45, 2003 [chapter 5].

[6] Philippon D, De Barros-Bouchet MI, Le Mogne T, Lerasle O, Bouffet A, Martin JM. Role of nascent metallic surfaces on the tribochemistry of phosphite lubricant additives. Tribol Intl 2011;44(6):684−91.

[7] Fujita H, Glovnea RP, Spikes HA. Study of zinc dialkyldithiophosphate antiwear film formation and removal processes, Part I: experimental. Tribol Trans 2005; 48(4):558—66.

[8] Archard JF. Wear theory and mechanisms. In: Peterson MB, Winer WO, editors. Wear control handbook. New York, NY: ASME; 1980. p. 35—80.

[9] Sullivan JL, Quinn TFJ, Rowson DM. Developments in the oxidational theory of mild wear. Tribol Intl 1980;13(4):153—8.

[10] Quinn TFJ. Review of oxidational wear, Part I: the origins of oxidational wear. Tribol Intl 1983;16(5):257—71.

[11] Quinn TFJ. Oxidational wear. In: Blau PJ, editor. ASM handbook, vol. 18: Friction, lubrication, and wear technology. ASM International, Materials Park, OH 1992. p. 280—9.

[12] Rowe CN. Lubricated wear. In: Peterson MB, Winer WO, editors. Wear control handbook. New York, NY: ASME; 1980. p. 143—60.

[13] Karmakar S, Rao URK, Sethuramiah A. Characterisation of sliding wear in dynamically stressed material. Wear B 1993;162:1081—90.

[14] Karmakar S, Rao URK, Sethuramiah A. An approach towards fatigue wear modelling. Wear 1996;198:242—50.

[15] Bosman R, Schipper DJ. Mild wear prediction of boundary-lubricated contacts. Tribol Lett 2011;42(2):169—78.

[16] Sankar PR, Kumar R, Sethuramiah A. Dry wear of grey cast iron. In: 2000 AIMETA international tribology conference, Sep 2000, L'Aquila, Italy, p. 127—35.

[17] Kumar R, Prakash B, Sethuramiah A. A systematic methodology to characterise the running-in and steady state wear process. Wear 2002;252(5—6):445—53.

[18] Quinn TFJ. Review of oxidational wear, Part II: recent developments and future trends in oxidational wear research. Tribol Intl 1983;16(6):305—15.

[19] Sakurai T, Ikeda S, Okabe H. The mechanism of reaction of sulphur compounds with steel surface during boundary lubrication using S^{35} as a tracer. ASLE Trans 1962;5:67—74.

[20] Kloss H, Wasche R. Analytical approach for wear prediction of metallic and ceramic materials in tribological applications. Wear 2009;266(3—4):476—81.

[21] Fillot N, Lordanoff I, Berthier Y. Simulation of wear through mass balance in a dry contact. Trans ASME J Tribol 2005;127(1):230—7.

[22] Heinicke G. Tribochemistry. Berlin: Akademie Verlag; 1984.

[23] Bulgarevich SB, Boiko MV, Tarasova EN, Feizova VA, Lebedinskii KS. Kinetics of mechanoactivation of tribochemical processes. J Friction Wear 2012;33(5):345—53 Allerton Press (Translation from Russian).

[24] Fujita H, Spikes HA. Study of zinc dialkyldithiophosphate antiwear film formation and removal processes, Part II: kinetic model. Tribol Trans 2005;48(4):567—75.

[25] Liu E, Kouame SD. An XPS study on the composition of zinc dialkyl dithiophosphate tribofilms and their effect on camshaft lobe wear. Tribol Trans 2014;57(1):18—27.

[26] Ravi D, Sethuramiah A. Acoustic emission in dynamic compression and its relevance to tribology. Trib Intl 1995;28(2):301—6.

CHAPTER 6

Statistics and Experimental Design in Perspective

6.1 INTRODUCTION

Design of experiments (DOE) is a statistical and mathematical tool to perform the experiments in a systematic way and analyze the data efficiently. In DOE the levels of factors are changed simultaneously to find the effect of individual factors as well as their interactions on response. The major factors in the present context can be concentration of additives, load, roughness, and temperature. The response of interest may be running-in wear rate, steady state wear rate, running-in period, or load carrying capacity. In some cases multiple responses may be of interest. The significant effect of the variables on the response is quantified by doing variance analysis with reference to experimental error estimated by repeating experiments at the same level of factors. The experimental errors are inherent in experimentation. It is caused due to several reasons that include atmospheric conditions, material inhomogeneity, operators' variability in conducting experiments, measurement errors, and nonstandardization of test samples. If the mean variance of response due to a factor is sufficiently large compared to mean experimental error then that factor is called a significant factor. The mean variance is found by dividing the variance by degree of freedom, which will be discussed in the next section.

The development of DOE is well documented in literature [1,2]. It was first successfully implemented in the agriculture sector in the 1930s by Fisher [3], who has introduced the three main terms of randomization, replication, and blocking widely used in DOE. He has given the concept of analysis of variance (ANOVA), factorial, and fractional factorial design. The next development starts in 1950s when Box and Wilson [4] introduced the concept of response surface methodology (RSM), which was applied in process industries like chemical industries. In these industries data can be obtained quickly and the experimenter can get sufficient information from small runs, which can be used in planning the next set of experiments for optimization. In the 1980s, Taguchi [5] introduced

Modeling of Chemical Wear.
DOI: http://dx.doi.org/10.1016/B978-0-12-804533-6.00006-8
© 2016 Elsevier Inc.
All rights reserved. 129

robust parameter design, which has popularized the use of DOE in other industries like automotive, aerospace, and electronics. His approach is to develop the product or process that is robust enough to remain insensitive to the environment or other factors that are difficult to control.

The literature on DOE is vast. In this chapter only the portion relevant to chemical wear modeling is discussed. First, the statistical foundation of DOE is explained. While the statistical basis appears rather abstract its use is well demonstrated with a non-numeric example. This is followed by a consideration of factorial and fractional designs. Multiple linear regression models are then discussed.

The concepts of RSM and multiple objective optimization are not considered here; they are dealt within Chapter 8.

The principles of DOE have been applied in the next two chapters, which deal with wear modeling and additive interactions.

6.2 STATISTICAL FOUNDATION OF DOE

This section discusses the relevant basic statistical concepts, hypothesis testing, and optimum sample size that form the foundation for understanding the other sections of this chapter.

6.2.1 Random Variable

In a statistical sense any experimental observation y is subject to error that includes several components. These components include errors associated with process, measurement, and environment among others, and are uncontrollable in nature. In view of the variations involved it is usual to consider y as a random variable. The random variable y may be discrete or continuous. The distribution of all possible values of y; that is, its population is expressed by probability function $p(y)$ if it is discrete and by probability density function $f(y)$ if it is continuous. The population of y is statistically characterized mainly by mean μ and variance σ^2, which quantify its central tendency and variability, respectively. Mathematically they are expressed as the following:

For continuous random variable:

$$\mu = E(y) = \int_{-\infty}^{\infty} yf(y)dy \quad \text{and} \quad \sigma^2 = V(y) = \int_{-\infty}^{\infty} (y-\mu)^2 f(y)dy \quad (6.1)$$

For discrete random variable:

$$\mu = E(y) = \sum_{\text{all possible } y} yp(y) \text{ and } \sigma^2 = V(y) = \sum_{\text{all possible } y} (y-\mu)^2 p(y) \quad (6.2)$$

where $E(y)$ represents the expected value of y and V is the variance operator. Using this equation, V operator can be shown in terms of E operator as $V = E[(y-\mu)^2]$. Using these operators and following these equations it can be easily shown that if y_1 and y_2 are two random variables having means as μ_1 and μ_2, respectively, and variances as σ_1^2 and σ_2^2 then

$$E(y_1 - y_2) = \mu_1 - \mu_2 \text{ and } V(y_1 - y_2) = \sigma_1^2 + \sigma_2^2 - 2 \text{ Covariance}(y_1, y_2)$$
$$(6.3)$$

$$E(y_1 + y_2) = \mu_1 + \mu_2 \text{ and } V(y_1 + y_2) = \sigma_1^2 + \sigma_2^2 + 2 \text{ Covariance}(y_1, y_2)$$
$$(6.4)$$

where $\text{Covariance}(y_1, y_2) = E[(y_1 - \mu_1)(y_2 - \mu_2)]$. If y_1 and y_2 are independent random variables then $\text{Covariance}(y_1, y_2) = 0$.

6.2.2 Normal Distribution

Most of the random variables follow normal distribution $N(\mu, \sigma^2)$. The shape of normal distribution may be seen in Figure 6.1, which is a distribution of a statistic to be discussed later in Section 6.2.5. The notation $N(\mu, \sigma^2)$ denotes normal distribution with mean μ and variance σ^2. The location and shape of this distribution are determined by parameters μ and σ^2, respectively. Normal distribution follows the reproductive property for addition and subtraction operations. Thus if y_1 and y_2 are two independent random variables following normal distributions $N(\mu_1, \sigma_1^2)$ and $N(\mu_2, \sigma_2^2)$, respectively, then $y_1 - y_2$ and $y_1 + y_2$ will also follow normal distributions as $N(\mu_1 - \mu_2, \sigma_1^2 + \sigma_2^2)$ and $N(\mu_1 + \mu_2, \sigma_1^2 + \sigma_2^2)$, respectively. This reproductive property also holds for n number of independent normally distributed random variables [6,7].

It is better to express the random variable y following $N(\mu, \sigma^2)$ as a standard normal random variable z as

$$z = \frac{y - \mu}{\sigma} \quad (6.5)$$

The variable z follows $N(0,1)$. This standardization helps in using a cumulative standard normal distribution table for statistical tests [8,9].

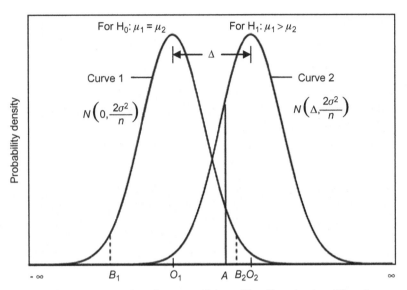

Difference in average of wear coefficients $\overline{y}_1 - \overline{y}_2$ using two different lubricants each having sample size of n

Figure 6.1 Probability density functions of difference in average $\overline{y}_1 - \overline{y}_2$ of wear coefficients with two different lubricants for two different hypotheses.

6.2.3 Sample Mean, Variance, and Degree of Freedom

The mean and variance estimation by using Eqns (6.1) and (6.2) need a very large number of observations and hence are impractical. So these are estimated by randomly taking a sample of an appropriate size n from the population. *Random sampling* means that each sample of a particular size has equal probability to be chosen. A function of sample observations having no unknown parameters is called a *statistic*. The mean and variance of the random sample is estimated by the statistics sample mean $\overline{y}$ and sample variance S^2. These statistics are unbiased estimators of μ and σ^2. The equations to compute $\overline{y}$ and S^2 are:

$$\overline{y} = \frac{\sum\limits_{i=1}^{n} y_i}{n} \qquad (6.6)$$

$$S^2 = \frac{SS}{n-1} = \frac{\text{Sum of squares}}{n-1} = \frac{\sum\limits_{i=1}^{n} (y_i - \overline{y})^2}{n-1} \qquad (6.7)$$

Applying E and V operators in Eqn (6.6), we get

$$E(\bar{y}) = \mu \quad \text{and} \quad V(\bar{y}) = \frac{\sigma^2}{n} \qquad (6.8)$$

It should also be noted that in Eqn (6.7) S^2 is estimated by dividing SS by $n - 1$ rather than n, while SS is obtained by adding n terms. The reason is that in SS, out of n terms only $n - 1$ terms are independent as $\sum_{i=1}^{n}(y_i - \bar{y}) = 0$. The number of independent terms $n-1$ in SS is known as *degree of freedom* (*df*) of SS. In any statistical quantity *df* can be computed by subtracting the number of statistics used to compute that quantity from the number of observations.

6.2.4 Sampling Distributions

The distribution of a statistic is known as sampling distribution. The sampling distribution of sample mean is an important distribution. It follows approximately the normal distribution even if the population from which the sample is drawn is not normal. This is due to the central limit theorem, which states that if we take a random sample of size n from a population that may be finite or infinite with mean μ and variance σ^2 then sample mean $\bar{y}$ will follow approximately $N(\mu, \sigma^2/n)$ if sample size is *large*. The central limit theorem allows the use of z_{01} and z_{02} statistics mentioned in Table 6.1 even if the samples are drawn from a non-normal population. Here it should be noted that if the population is normal then the $\bar{y}$ statistic will definitely follow $N(\mu, \sigma^2/n)$, which can be easily proved by using the reproductive property and Eqn (6.8). Sampling distribution of the important test statistics along with their tribological application are given in Table 6.1. The second and first columns of the table give information about the sample and its population. It should be noted here that samples should be drawn randomly in order to estimate a particular test statistic. The symbols used in the table that have not been described earlier are given at the end of table.

In Table 6.1 z_{01} and t_{01} statistics are used to compare the population mean μ with a fixed value μ_0. z_{02} and t_{02} are used to compare the means μ_1 and μ_2 of two populations. z statistic is used if the population variance is known. Otherwise t statistic should be used. t_{02} statistic has an assumption that $\sigma_1^2 = \sigma_2^2$ though the value is not known. The estimation of variability needed for calculation is obtained from the pooled value of both

Table 6.1 Sampling distributions of important test statistics with details of random sampling from population(s) along with applications

Population(s)	Sample(s)	Test statistic	Sampling distribution	Tribological examples for use of test statistic
Single population: $N(\mu, \sigma^2)$	One random sample of size n	$z_{01} = \dfrac{\bar{y} - \mu_0}{\sigma/\sqrt{n}}$	$N(0,1)$	Compare the new lubricant for load carrying capacity and wear coefficient with desired values
		$t_{01} = \dfrac{\bar{y} - \mu_0}{S/\sqrt{n}}$	t_{n-1}	
		$\chi_0^2 = \dfrac{(n-1)S^2}{\sigma_0^2}$	χ_{n-1}^2	Assess whether the old wear testing machine is giving the precision as prescribed
Two independent populations: $N(\mu_1, \sigma_1^2)$, $N(\mu_2, \sigma_2^2)$	Two random samples: one sample of size n_1 from population 1 and another of size n_2 from population 2	$z_{02} = \dfrac{\bar{y}_1 - \bar{y}_2}{\sqrt{\dfrac{\sigma_1^2}{n_1} + \dfrac{\sigma_2^2}{n_2}}}$	$N(0,1)$	To compare two different lubricants on the basis of a. wear coefficient b. load carrying capacity c. oxidation stability
		$t_{02} = \dfrac{\bar{y}_1 - \bar{y}_2}{S_p\sqrt{\dfrac{1}{n_1} + \dfrac{1}{n_2}}}$	$t_{n_1+n_2-2}$	
		$F_0 = \dfrac{S_1^2}{S_2^2}$	F_{n_1-1,n_2-1}	Analysis of a. designed experiment of wear test b. regression wear model

where

S_p = Square root of pooled sample variance = $\sqrt{\dfrac{\text{Total SS}}{df}} = \sqrt{\dfrac{SS_1 + SS_2}{(n_1-1)+(n_2-1)}} = \sqrt{\dfrac{(n_1-1)S_1^2 + (n_2-1)S_2^2}{n_1+n_2-2}}$

t_{n-1} = t-distribution with $n-1$ df,

$t_{n_1+n_2-2}$ = t-distribution with $n_1 + n_2 - 2$ df,

χ_{n-1}^2 = Chi-square distribution with $n-1$ df,

F_{n_1-1,n_2-1} = F-distribution with numerator df of $n_1 - 1$ and denominator df of $n_2 - 1$.

the samples S_p as given in Table 6.1. If this condition is not satisfied then another statistic given in literature should be used [10]. χ_0^2 statistic is used to compare population variance σ^2 with a fixed value σ_0^2. F_0 statistic is used to compare the variances σ_1^2 and σ_2^2 of two populations. All these test statistics are based on standard random variables. The detailed procedure of development and use of the test statistic z_{02} are exemplified in hypothesis testing next.

6.2.5 Hypothesis Testing, Errors in Testing, and Optimum Sample Size Selection

The hypothesis of relevance here is based on statistical comparison of μ and σ^2 of one or more populations. Hypothesis testing is the decision-making method to accept or reject the statement. The statement to be tested is called *null hypothesis* denoted by H_0. Along with null hypothesis an *alternative hypothesis* denoted by H_1 is also specified, which gives the condition to be concluded at rejection. Alternative hypotheses are of two types: (1) *one-sided alternative hypothesis* and (2) *two-sided alternative hypothesis*. These are discussed later in a non-numeric example. The hypothesis testing involves these steps:

1. Experimentation with a randomized sample.
2. Selection of an appropriate test statistic that follows certain sampling distribution.
3. Computation of the test statistic.
4. Selection of critical or rejection region in sampling distribution of interest.
5. Decision on the null hypothesis.
   In hypothesis testing two types of errors are encountered:
1. Type-I error: H_0 is rejected when it was true.
2. Type-II error: H_0 is accepted when it was false.
   Probability of these errors is denoted by α and β, which are expressed as
   $\alpha = P(\text{Type-I error})$,
   $\beta = P(\text{Type-II error})$.
   These probabilities are discussed at the end of the following example.
   Let us consider a non-numeric example in which an oil lubricant formulator wants to change the existing engine oil, lubricant 1, with another eco-friendly lubricant, lubricant 2. He wants to know first whether the two lubricants are different in terms of engine wear. In case they are different, he would additionally like to know if the eco-friendly lubricant is superior. To get these answers tests on real engines at the same operating

conditions can be performed and wear coefficient can be estimated by the methodology developed by the authors [11]. Suppose n engine tests have been decided with each lubricant. Total $2n$ tests should be conducted in a random order. It is assumed that variance σ^2 in both type of tests is the same and its value is known from experience.

For this problem two separate hypothesis tests are done. In both tests the null hypothesis is H_0: $\mu_1 = \mu_2$; that is, $\mu_1 - \mu_2 = 0$, which expresses that both populations 1 and 2 are same. To test this null hypothesis distribution of $\bar{y}_1 - \bar{y}_2$ has to be observed where $\bar{y}_1$ and $\bar{y}_2$ are the respective averages of n wear coefficients found with lubricants 1 and 2, respectively. If populations 1 and 2 are independent normal distributions then using Eqns (6.8) and (6.3), and the reproductive property of normal distribution it can be said that $\bar{y}_1 - \bar{y}_2$ will also be normally distributed as $N(\mu_1 - \mu_2, \sigma_1^2/n + \sigma_2^2/n)$. Under null hypothesis H_0: $\mu_1 - \mu_2 = 0$ this distribution will be $N(0, \sigma_1^2/n + \sigma_2^2/n)$, which has been shown by curve 1 in Figure 6.1. This curve varies from $-\infty$ to $+\infty$ and is centered at O_1. Now using Eqn (6.5) the appropriate test statistic for H_0: $\mu_1 - \mu_2 = 0$ can be developed as test statistics of a standard random variable. The developed statistic (Table 6.1) is given as

$$z_{02} = (\bar{y}_1 - \bar{y}_2)/\sqrt{\sigma_1^2/n + \sigma_2^2/n} \tag{6.9}$$

This statistic follows the standard normal distribution $N(0,1)$.

Since for the above problem $\sigma_1^2 = \sigma_2^2 = \sigma^2$ is assumed the test statistic can be rewritten as

$$z_{02} = \frac{\bar{y}_1 - \bar{y}_2}{\sigma}\sqrt{\frac{n}{2}} \tag{6.10}$$

If σ_1^2 and σ_2^2 are not known but $\sigma_1^2 = \sigma_2^2 = \sigma^2$ can be reasonably assumed then it is estimated by pooled sample variance

$$S_p^2 = \frac{\text{Total SS}}{df} = \frac{SS_1 + SS_2}{2(n-1)}$$

Using Eqn (6.7) we can get $S_p^2 = (S_1^2 + S_2^2)/2$ and the test statistic will be $t_{02} = ((\bar{y}_1 - \bar{y}_2)/S_p)\sqrt{n/2}$, which follows t-distribution with $2(n-1)$ df.

For the first hypothesis test an alternative hypothesis is H_1: $\mu_1 \neq \mu_2$. This is a *two-sided alternative hypothesis* in which the significance level of α, the critical region, is in both tail sides of distribution. In Figure 6.1 it is shown under curve 1 from $-\infty$ to B_1 and from B_2 to ∞. The area under the curve 1 in these critical regions in both sides is $\alpha/2$. The critical regions for this case are obtained using the cumulative standard normal distribution table, which gives the area under $N(0,1)$ curve from $-\infty$ to z. From this table we find the z value, which gives area $= \alpha/2$ and denote $|z|$ as $z_{\alpha/2}$. Now calculate z_{02} using Eqn (6.10). If $|z_{02}| > z_{\alpha/2}$ (i.e., z_{02} falls in a critical region), then the null hypothesis H_0 is rejected and we say that both lubricants are really different in their wear response.

For the second hypothesis test an alternative hypothesis is H_1: $\mu_1 > \mu_2$. This is a *one-sided alternative hypothesis* in which the significance level of α, the critical region, is in one tail side of distribution. In Figure 6.1 it is shown under curve 1 from A to ∞. The area under the curve 1 in this critical region is α. If $|z_{02}|$ is found to be greater than z_α then null hypothesis H_0 is rejected and we say that lubricant 1 is giving higher wear coefficient than lubricant 2. In other words we can say that lubricant 2, the eco-friendly lubricant, has better antiwear property than existing engine oil.

In these conclusions type-I and type-II errors are encountered as mentioned earlier. In the type-I error, the fact is H_0: $\mu_1 - \mu_2 = 0$, hence curve 1 in Figure 6.1 should be referred for this error. The probability of making the wrong decision of rejecting H_0 will be the area of the critical region, which is α. In the type-II error, H_0 is false, hence distribution of $\bar{y}_1 - \bar{y}_2$ shown by curve 1 is not correct. Let the correct distribution be $N(\Delta, \sigma_1^2/n + \sigma_2^2/n)$ as shown by curve 2, which is centered at O_2. Δ is the difference between the means of true and hypothesized distribution of $\bar{y}_1 - \bar{y}_2$. The probability of making the wrong decision of accepting H_0 will be the area of acceptance region under curve 2, which is denoted by β. For the first part of the problem, β is the area under curve 2 from B_1 to B_2; for second part it is from $-\infty$ to A. Referring to Figure 6.1 we can easily draw the following conclusions:

1. β value depends on choice of α value. With the increase of α the value of β decreases and vice versa.
2. Increasing the Δ difference between means of true and hypothesized distribution the β value decreases.
3. Increasing the sample size will decrease the variability of distribution as $V(\bar{y}_1 - \bar{y}_2) = \sigma_1^2/n + \sigma_2^2/n$. Hence thinner distribution curves will be obtained, which will reduce both α and β.

From this discussion it is clear that an optimum sample size selection depends on α and β values. Let us find the sample size for part II of the problem. Referring to Figure 6.1 if we convert the $\bar{y}_1 - \bar{y}_2$ random variable into the standard normal random variable with respect to H_0: $\mu_1 - \mu_2 = 0$ the curves 1 and 2 will be distributed as $N(0,1)$ and $N((\Delta/\sigma)\sqrt{n/2},1)$, respectively. The x-axis will give z_{02} value. The x-coordinate of A will be z_α. Converting again z_{02} into a standard normal random variable with respect to distribution $N((\Delta/\sigma)\sqrt{n/2}, 1)$ will give the x-coordinate of A as $z_\alpha - (\Delta/\sigma)\sqrt{n/2}$ and curve 2 will be distributed as $N(0,1)$. The area under this distribution from $-\infty$ to $-z_\beta$ will be β, which is actually from $-\infty$ to A. Hence we can write the following relationship:

$$-z_\beta = z_\alpha - \frac{\Delta}{\sigma}\sqrt{\frac{n}{2}} \qquad (6.11)$$

Rearranging Eqn (6.11) we get

$$n = \frac{(z_\alpha + z_\beta)^2 2\sigma^2}{\Delta^2} \qquad (6.12)$$

This equation can be used in selecting an appropriate sample size.

Another approach to find the sample size is the use of operating characteristic (OC) curves. The appropriate set of OC curves is chosen from different sets of OC curves available in the handbook [8] or in [1,10] on the basis of significance level α, sampling distribution, and type of alternative hypothesis. The appropriate set of OC curves chosen for the second part of the problem has many curves for different sample size. These curves show β variation as a function of d a nondimensional difference between means of true and hypothesized distribution. The d is given as

$$d = \frac{\Delta}{\sqrt{\sigma_1^2 + \sigma_2^2}} \qquad (6.13)$$

By knowing the value of Δ, variance, α, and β values the proper sample size may be selected.

Software with different levels of sophistication is available for statistical analysis as well as DOE; SAS and Design-Expert are widely used in these areas.

6.3 MAIN CONSIDERATIONS IN EXPERIMENTAL DESIGN

The DOE relies heavily on the statistical principles discussed in previous section. At the outset it is necessary to emphasize the statistical principles involved in a consolidated manner. Two basic principles involved are randomization and replication. Also blocking may be considered as an additional principle involved.

Randomization is a key issue in a designed experiment, which makes sure that responses and errors are independently distributed random variables. This is one of the important assumptions for statistical methods. Randomization should be done at each stage of the experiment with regard to test specimens, lubricant formulations, and testing order. It evenly distributes the experimental error in design space leading to unbiased results. Random order can be obtained either by using random number tables or software.

Replication means performing the experiment many times independently at selected experimental design points. Replication may be considered as a repetition that is made in random order. For example, consider five experiments repeated twice at five selected conditions involving a total of 10 experiments. If we randomize all 10 tests the repetitions are done in random order and not one after the other. It differs from repetition, in which the tests are done one after the other at the given condition. For example, hardness measurement at a given condition three times is a repetition. Here the same test is done three times in sequence. Replication serves mainly two purposes:

1. Estimation of experimental error on which analysis of the designed experiment is based.
2. Increases the ability to differentiate between the influence of various factors due to increase in sample size as discussed in an earlier section.

Blocking may be considered as a technique that reduces the influence of nuisance factors. For example, if wear comparisons are done between two lubricants in a reciprocating tester roughness variations in test pieces may influence the performance. On the other hand if we restrict the roughness to a specific value, the two lubricants can be clearly distinguished. In effect we have blocked a factor—this is referred to as blocking. Roughness is a nuisance factor because in our study we are not interested in its influence on wear. In a broader sense blocking involves a special methodology to minimize the influence of the nuisance factor. This makes comparison between interested factors more clear. Let us consider the same example in which we want to know whether two lubricants are different in terms of wear by conducting experiments in a reciprocating rig. If the roughness of different lower test specimens varies too much it will create a noise in comparing lubricants as roughness affects wear [12]. If we treat each specimen as a block and conduct two tests in a reciprocating rig with both lubricants on the same specimen at different locations in a random way the variability due to roughness will be reduced.

For analysis, the t_{01} test statistic mentioned in Table 6.1 may be used. It can be computed by recording data as $d_i = y_{1i} - y_{2i}$, where y_{1i} and y_{2i} are the wear coefficients due to lubricants 1 and 2 of the ith specimen. The null and alternative hypotheses can be taken as H_0: $\mu_d = 0$ and H_1: $\mu_d \neq 0$, respectively. If we want to compare more than two lubricants, a randomized complete block design should be used, which has not been discussed in the book. The details of using blocking in experimental design may be obtained from the literature [1,2,13].

In blocking, randomization should be done within the block. Also blocking is effective only if the variability within the block is less than between blocks. If this condition is not satisfied then blocking will be the wrong choice and the test should be conducted in a completely random way in the entire design space.

6.4 FACTORIAL DESIGN

This section first compares a one-factor-at-a-time approach with factorial design. The coding formulae for levels of a factor are also given, which are normally used in modeling. ANOVA is then discussed.

6.4.1 Comparison Between One-Factor-at-a-Time and Factorial Design Experiments

In the usual one-factor-at-a-time-approach, one factor is varied while other factors are kept constant. In DOE factorial design is used so that all possible combinations of the levels of the factors are investigated in the experiment. Figure 6.2 shows the comparison between two approaches. In this figure the coded value of five levels for concentrations of two additives ZDDP and MoDTC are shown in horizontal and vertical axes. The five levels are -1.4, -1.0, 0, 1.0, and 1.4. The coding values in this example can be decoded in % (w/v) from this formula:

$$x_i = \frac{z_i - z_0}{\Delta z} \tag{6.14}$$

where

x_i = coded value of additive concentration at ith experiment,

z_i = additive concentration in % (w/v) at ith experiment,

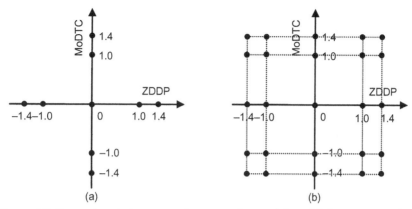

Figure 6.2 Comparison between one-factor-at-a-time and factorial design approaches. (a) Experimental points for two factors with five levels in one-factor-at-a-time experiment, (b) experimental points for two factors with five levels in factorial design experiment.

z_0 = the additive concentration in % (w/v) at mid level (i.e., at the center point),

$$\Delta z = \text{the step value} = \frac{z_{max} - z_{min}}{x_{max} - x_{min}} = \frac{z_{max} - z_{min}}{1.4 - (-1.4)} = \frac{z_{max} - z_{min}}{2.8},$$

z_{max}, z_{min} = maximum and minimum values of additive concentrations in % (w/v),

x_{max}, x_{min} = maximum and minimum values of additive concentrations in coded form.

In the above, coding variation in the range of levels is linear. The nonlinear variation can also be taken into account. One form is [14,15]:

$$x_i = \frac{\log_e z_i - \log_e z_0}{\Delta z} \quad \text{where } \Delta z = \frac{\log_e z_{max} - \log_e z_{min}}{x_{max} - x_{min}} \quad (6.15)$$

Figure 6.2a depicts the one-factor-at-a-time method. Using this method the output response of wear is observed by varying concentration of one additive, keeping concentration of the other additive constant at central level. A total of nine experiments have to be performed. The solid circular points in the figure are the experimental points. For both additives if the experimenter wants to observe the variation in wear five times in the entire range the total number of experiments will be 45. Now suppose 5^2 full factorial design is used, in which the number of experiments to be performed by factorial deign = (level)^{factor}. Therefore, the number of experiments to be performed is $5^2 = 25$. Figure 6.2b shows the above design. From the figure we can easily see that we can obtain variation in wear five times by varying the concentration of one additive from -1.4 to 1.4 while keeping the concentration of other additive constant at a different level of -1.4, -1, 0, 1, and 1.4. In the second approach we have to perform only 25 experiments compared to 45 in the first approach for the same objective. The second approach also has the advantage of studying the interaction of factors. In the present example we can see whether there is synergy between two additives in terms of antiwear property.

6.4.2 ANOVA for Factorial Design

The ANOVA is used to quantify the significance of factors and their interactions. In order to understand the concept behind it, consider the example of 5^2 factorial design given earlier with two times repetition of experiments at all 25 combinations of levels of two factors. A total of 50 experiments have to be performed in a randomized way so that experimental error is uniformly distributed in the design space. Table 6.2 shows the wear rate w_{ijk} at different combinations of factors' level and replication. The nomenclature of symbols used here are given below Table 6.2.

The variability in wear rates in the entire 50 experiments quantified by total sum of squares SS_T expressed in Eqn (6.16) is due to various influences. These are the concentration effects of ZDDP and MoDTC, their interactions, and experimental errors.

$$\text{Total sum of squares } SS_T = \sum_{i=1}^{l_1}\sum_{j=1}^{l_2}\sum_{k=1}^{n}(w_{ijk}-\overline{w})^2 \qquad (6.16)$$

SS_T may be expressed as the summation of sum of squares (SS) due to various factors mentioned earlier [1,13]. It is the decomposition of total variability into various components, which is expressed as

$$SS_T = SS_1 + SS_2 + SS_{interaction} + SS_E \qquad (6.17)$$

This decomposition can also be expressed as summation of SS due to the combined effect of factor 1, factor 2, their interaction, and experimental error.

$$SS_T = SS_{1+2+interaction} + SS_E \qquad (6.18)$$

where

$SS_1 =$ sum of squares due to factor 1: ZDDP $= l_2 n \sum_{i=1}^{l_1}(\overline{w}_{f_1 i}-\overline{w})^2$,

$SS_2 =$ sum of squares due to factor 2: MoDTC $= l_1 n \sum_{j=1}^{l_2}(\overline{w}_{f_2 j}-\overline{w})^2$,

Table 6.2 Two-factor 5^2 factorial design having two replications with wear rate as response

Factor 1: ZDDP concentration (coded value)	Factor 2: MoDTC concentration (coded value)					Sum of wear rates in row	Average wear rate in row
	−1.4	−1.0	0	1.0	1.4		
−1.4	u_{111} u_{112} $\bar{u}_{11}$	u_{121} u_{122} $\bar{u}_{12}$	u_{131} u_{132} $\bar{u}_{13}$	u_{141} u_{142} $\bar{u}_{14}$	u_{151} u_{152} $\bar{u}_{15}$	$W_{f_1 1}$	$\bar{u}_{f_1 1}$
−1.0	u_{211} u_{212} $\bar{u}_{21}$	u_{221} u_{222} $\bar{u}_{22}$	u_{231} u_{232} $\bar{u}_{23}$	u_{241} u_{242} $\bar{u}_{24}$	u_{251} u_{252} $\bar{u}_{25}$	$W_{f_1 2}$	$\bar{u}_{f_1 2}$
0	u_{311} u_{312} $\bar{u}_{31}$	u_{321} u_{322} $\bar{u}_{32}$	u_{331} u_{332} $\bar{u}_{33}$	u_{341} u_{342} $\bar{u}_{34}$	u_{351} u_{352} $\bar{u}_{35}$	$W_{f_1 3}$	$\bar{u}_{f_1 3}$
1.0	u_{411} u_{412} $\bar{u}_{41}$	u_{421} u_{422} $\bar{u}_{42}$	u_{431} u_{432} $\bar{u}_{43}$	u_{441} u_{442} $\bar{u}_{44}$	u_{451} u_{452} $\bar{u}_{45}$	$W_{f_1 4}$	$\bar{u}_{f_1 4}$
1.4	u_{511} u_{512} $\bar{u}_{51}$	u_{521} u_{522} $\bar{u}_{52}$	u_{531} u_{532} $\bar{u}_{53}$	u_{541} u_{542} $\bar{u}_{54}$	u_{551} u_{552} $\bar{u}_{55}$	$W_{f_1 5}$	$\bar{u}_{f_1 5}$
Sum of wear rates in column	$W_{f_2 1}$	$W_{f_2 2}$	$W_{f_2 3}$	$W_{f_2 4}$	$W_{f_2 5}$	Overall sum of wear rate = W	
Average wear rate in column	$\bar{u}_{f_2 1}$	$\bar{u}_{f_2 2}$	$\bar{u}_{f_2 3}$	$\bar{u}_{f_2 4}$	$\bar{u}_{f_2 5}$		Overall average wear rate = $\bar{w}$

u_{ijk} = Wear rate at ith level of factor 1, jth level of factor 2, and kth replication. $1 \le i \le l_1$; $1 \le j \le l_2$; $1 \le k \le n$.

l_1 = Number of levels in factor 1. In the present example it is 5.

l_2 = Number of levels in factor 2. In the present example it is 5.

n = Number of replications. In the present example it is 2.

$\bar{u}_{ij}$ = Average wear rate at ijth cell. It represents the expected wear rate due to the combined effect of factor 1, factor 2, and interaction at ijth combination of level of factors in replicated space.

$W_{f_1 i}$ = Sum of wear rates in rth row = $\sum_j^n \sum_k^n u_{ijk}$.

$W_{f_2 j}$ = Sum of wear rates in jth column = $\sum_i^{l_1} \sum_k^n u_{ijk}$.

$\bar{u}_{f_1 i}$ = Average wear rate in rth row = $W_{f_1 i}/(l_2 n)$ representing the expected wear rate due to factor 1 at ith level in design space of factor 2.

$\bar{u}_{f_2 j}$ = Average wear rate in jth column = $W_{f_2 j}/(l_1 n)$ representing the expected wear rate due to factor 2 at jth level in design space of factor 1.

W = Overall sum of wear rates = $\sum_{i=1}^{l_1} \sum_{j=1}^{l_2} \sum_{k=1}^{n} u_{ijk}$.

$\bar{w}$ = Overall average wear rate = $W/(l_1 l_2 n)$. It represents expected wear rate due to factor 1, factor 2, interaction, and experimental error in entire design space.

Table 6.3 ANOVA table for wear rate data of two-factor 5^2 factorial design

Source	Sum of squares	Degree of freedom	Mean squares	F-ratio
Factor 1: ZDDP	SS_1	$l_1 - 1$	MS_1	MS_1/MS_E
Factor 2: MoDTC	SS_2	$l_2 - 1$	MS_2	MS_2/MS_E
Interaction	$SS_{interaction}$	$(l_1 l_2 - 1) - (l_1 - 1) - (l_2 - 1)$ $= (l_1 - 1)(l_2 - 1)$	$MS_{interaction}$	$MS_{interaction}/MS_E$
Error	SS_E	$l_1 l_2(n - 1)$	MS_E	
Total	SS_T	$l_1 l_2 n - 1$		

$SS_{1+2+interaction} = $ sum of squares due to factor 1, factor 2, and interaction $= n \sum_{i=1}^{l_1} \sum_{j=1}^{l_2} (\overline{w}_{ij} - \overline{w})^2$,

$SS_{interaction} = $ sum of squares due to interaction between factors $= SS_{1+2+interaction} - SS_1 - SS_2$,

$SS_E = $ sum of squares due to experimental error $= \sum_{i=1}^{l_1} \sum_{j=1}^{l_2} \sum_{k=1}^{n} (w_{ijk} - \overline{w}_{ij})^2$.

Now the significance test can be done by generating ANOVA (Table 6.3). In this table the mean squares are obtained by dividing SS by degree of freedom. It estimates the variance of source. In ANOVA the variance of factor is compared with the variance of experimental error. If the F-ratio is greater than F_{α,ν_1,ν_2} obtained from the F-distribution table available in statistical handbooks the factor is called significant. The α value is the level of significance, ν_1 and ν_2 are degree of freedom of factor and experimental error, respectively. The level of significance α is normally taken as 0.05, which signifies that the chances of error in conclusion of significance is 5%.

6.5 FRACTIONAL FACTORIAL DESIGN

The full factorial design is not economically viable if the number of factors increases. In that case fractional factorial design is used, which is a fraction of full factorial design. The scheme to obtain the design and the inherent aliasing of factorial effect are discussed below. The last subsection discusses the dealiasing techniques and an appropriate methodology for selection of fractional design.

6.5.1 Fractional Factorial Design: Two Levels

Two-level factorial design is expressed as $+1$ and -1 levels. The total number of experiments is 2^k, where k is the number of factors. As the number of factors increases the number of experiments increases rapidly.

It should be noted here that out of $2^k - 1$ degree of freedom (df) only $^kC_1 = k!/(1!(k-1)!)$ and $^kC_2 = k!/(2!(k-1)!)$ df are used for main and two-factor interactions. Other df obtained from doing large numbers of experiments is for higher order interaction terms that are normally insignificant. Hence fractional factorial design can be constructed so that expected levels of interaction only are included.

The two-level fractional design is constructed by choosing proper generator/s $I = ab\cdots c$, which has all levels of $+1$ or -1. $ab\cdots c$ represents multiplication of the corresponding level of three or more factors a, b, $\ldots$, and c. The examples shown in this section have generator/s of $+1$ levels. In $1/2^p$th fraction of 2^k design, which is expressed as 2^{k-p}, p independent generators are required to construct 2^{k-p} design. Out of these p generators the generator I_{smallest} having smallest number of factors defines the resolution of the design. The number of factors in I_{smallest}, $n(I_{\text{smallest}})$ is called as resolution of the design expressed in roman numbers.

Let us consider 2^{5-2} design shown in Table 6.4. This design has eight combinations of levels. The first three factors have levels according to 2^3 full factorial design; the fourth factor's levels are found by multiplying the first and second columns, and the fifth factor's levels are found by multiplying the first and third columns. The fourth and fifth columns' generation in short may be expressed in equation form as $4 = 12$ and $5 = 13$. It gives two generators $I = 124$ and $I = 135$, respectively, by multiplying 4 on both sides of the first relation $4 = 12$, and 5 on both sides of the second relation $5 = 13$, as 4^2 and 5^2 will give level $+1$ level. This design is called the resolution III design since both generators in this design have same number of factors 3, so $n(I_{\text{smallest}}) = 3$.

Table 6.4 Construction of 1/4th fraction of 2^5 design (i.e., 2^{5-2} design)

S. No.	Factor 1	Factor 2	Factor 3	Factor 4 $I = 124$ $4 = 12$	Factor 5 $I = 135$ $5 = 13$
1	−1	−1	−1	+1	+1
2	+1	−1	−1	−1	−1
3	−1	+1	−1	−1	+1
4	+1	+1	−1	+1	−1
5	−1	−1	+1	+1	−1
6	+1	−1	+1	−1	+1
7	−1	+1	+1	−1	−1
8	+1	+1	+1	+1	+1

From the relations $4 = 12$ and $5 = 13$ we see that it will be difficult to distinguish between the main effect of 4 and the interaction effect between 1 and 2. Similarly it will be difficult to find whether the behavior in response is due to factor 5 or due to the interaction between 1 and 3. Factor 4 is said to be aliased with 12 interactions and factor 5 is aliased with 13 interactions. The word alias refers to the situation where interaction between a given number of factors gets confounded with other interactions or main factors. All aliases for this design can be obtained from the above two generators and are given below.

$1 = 24 = 35 = 12345$, $2 = 14 = 345 = 1235$, $3 = 15 = 245 = 1234$,
$4 = 12 = 235 = 1345$, $5 = 13 = 234 = 1245$, $23 = 45 = 125 = 134$,
$25 = 34 = 123 = 145$.

From these aliases we can say that the resolution III design is suitable only if the interest is to find the main factor effects. This is true provided it is known that two-factor and higher order interactions are insignificant. Other resolution III designs can be generated by choosing any two independent generators from six generators: $I = 124 = 134 = 234 = 125 = 135 = 235$.

The ANOVA for the above design can be obtained from the procedure given in the previous section in which seven effects: $1 + 24 + 35 + 12345$, $2 + 14 + 345 + 1235$, $3 + 15 + 245 + 1234$, $4 + 12 + 235 + 1345$, $5 + 13 + 234 + 1245$, $23 + 45 + 125 + 134$, $25 + 34 + 123 + 145$, each having one df is estimated. Of course as stated earlier, to do ANOVA the above design must have ≥ 2 replications to estimate experimental error.

If the experimenters' interest is to find the main factor effect with confidence then the resolution IV design is useful. In this design main factors are not aliased with two-factor interaction. The example is 2^{4-1} design in which the $I = 1234$ generator is used to get the fourth factor's levels. ANOVA of this design will estimate the following seven effects in which each has one df: $1 + 234$, $2 + 134$, $3 + 124$, $4 + 123$, $12 + 34$, $13 + 24$, and $14 + 23$. From these aliases we can say that if we have knowledge that three-factor interactions are insignificant as normally happens, main factor effects can be investigated clearly.

To find the main factor and two-factor interaction effect without any ambiguity the resolution V design is useful. In this design the main factor and two-factor interactions are not aliased with any main factor and two-factor interaction. The example is 2^{5-1} design in which the $I = 12345$ generator is used to get levels of the fifth factor. ANOVA of this design will estimate the following 15 effects each having one df: $1 + 2345$, $2 + 1345$, $3 + 1245$, $4 + 1235$, $5 + 1234$, $12 + 345$, $13 + 245$, $14 + 235$, $15 + 234$, $23 + 145$, $24 + 135$, $25 + 134$, $34 + 123$, $35 + 124$, $45 + 123$. These aliases clearly indicate that if higher order ≥ 3 interactions are insignificant the main and two-factor interaction effect can be estimated with confidence.

6.5.2 Fractional Factorial Design: Three Levels

Three-level factorial design allows taking into account the response curvature if it exists in modeling. Three levels in this design are usually represented numerically by designating low, medium, and high levels as 0, 1, and 2. Fractional design in this case is developed by taking generator I as the summation of the level of factors. Sum value, s_{m3}, in the generator equation is represented as the remainder of Sum/3. This is called modulus 3 (mod 3). The value of s_{m3} may be 0, 1, or 2. Let us consider 1/3 fraction of 3^3 factorial design represented as 3^{3-1} design. Let the factors be represented by A, B, and C. The generator I can be chosen from any four orthogonal components of three-factor interactions: ABC, AB^2C, ABC^2, and AB^2C^2. The respective generators are $x_1 + x_2 + x_3 = s_{m3}$, $x_1 + 2x_2 + x_3 = s_{m3}$, $x_1 + x_2 + 2x_3 = s_{m3}$, and $x_1 + 2x_2 + 2x_3 = s_{m3}$. Exponent 2 in B and C represents two times addition in the generator equation and hence are expressed as coefficients in the respective factors' level x in the generator equation.

Since s_{m3} can be 0, 1, or 2, a total of 12 different 3^{3-1} designs can be generated. Let us select the last generator with $s_{m3} = 1$. In 3^{3-1} design the first two factors A and B have levels according to 3^2 full factorial design. The levels of the third factor C are found from equation $x_1 + 2x_2 + 2x_3 = 1$ (mod 3). Because s_{m3} represents modulus 3, a multiple of 3 is added in the right-hand side of the equation whenever required in order to find the level of the third factor. Following this procedure the nine different combinations of levels found are 002, 011, 020, 100, 112, 121, 201, 210, and 222.

All aliases of this design can be obtained from I and I^2. ANOVA of this design estimates the following four effects each having two df: $A + BC + ABC$, $B + AC^2 + ABC^2$, $C + AB^2 + AB^2C$, and $AB + AC + BC^2$ [1]. Care should be taken in analyzing the main effect with these aliases. If the factors are of an independent nature and do not interact with other factors this design is suitable to investigate the effect of factors with a possible nonlinear nature of response. The modeling of response curvature can also be done by using RSM. The selection of three-level design has been discussed in detail in Chapter 7.

6.5.3 Dealiasing and Selection of Optimum Fractional Design

The previous subsections on two- and three-level factorial designs did consider the issue of selection between alternate designs. The present subsection provides a more formal methodology.

The best fractional design selection and ambiguity removal of aliased terms becomes easier if the following empirical principles are considered: hierarchy, sparsity, and heredity. The hierarchy principle states that the effect of factors having lower order has more probability to be significant than of higher order. It suggests that if resources are limited then priority should be given to the lower order factor. The sparsity principle indicates that out of several variables the process is likely to be affected by a few factors only. The heredity principle says that interaction terms can be significant only if at least one of its parent factors is significant [16].

6.5.3.1 Dealiasing Technique

The ambiguity between aliased factors can be removed by following these considerations:

1. If factors of different orders are aliased then using the hierarchy principle, higher order factors may be ignored in comparison to lower order.
2. If the aliased factors are of the same order then prior knowledge of nonsignificant factors may dealiase the effect.
3. In a study of the whole process the factors can be grouped into different subprocesses. Taking wear process as an example, lubricant formulation, surface formation, and operating conditions are the subprocesses. The factor interaction within subprocesses has more probability to be significant than factor interaction between subprocesses.
4. More experiments other than the original can be planned that provide additional information to dispel the confusion among aliased factors. The details of conducting such experiments are available in literature [2].

6.5.3.2 Optimum Selection of Fractional Design

Selection of an optimum fraction is normally obtained by using two criteria, namely maximum resolution [17,18] and minimum aberration [19]. In order to understand these criteria let us define a pattern vector

$W = (A_3, A_4, \ldots, A_k)$ for 2^{k-p} design. In this vector A_i is the number of generators having i factors. The smallest r, such that $A_r \geq 1$, is known as resolution of a 2^{k-p} design. The maximum resolution criterion selects the fractional design, which has maximum resolution. Minimum aberration criterion states that for any two 2^{k-p} designs d_1 and d_2, let r be the smallest integer such that $A_r(d_1) \neq A_r(d_2)$. Then d_1 is said to have less aberration than d_2 if $A_r(d_1) < A_r(d_2)$. The reason is that large number of generators having r factors will give more aliases, hence more aberration. If there is no design with less aberration than d_1 then d_1 has minimum aberration [2].

The best fractional design depends mainly on the objective of experiments. Table 6.5 of some important general objectives, recommendations, and its limitations may help in selection and design of fractional experiments.

Table 6.5 Recommendation and limitation to select a fractional design for different objectives

Objectives	Recommendation to select fractional design	Limitation
Linear modeling of response	Designs having two levels	
Nonlinear modeling of response	Designs having ≥ 3 levels	
Screening of main factors	Design having III resolution	Factors having ≥ 2 order are insignificant[a]
Clear effect of main factors	Design having IV resolution[b]	Factors having ≥ 3 order are insignificant[c]
Clear effect of main factors and two-factor interactions	Design having V resolution	Factors having ≥ 3 order are insignificant[c]

[a]Some 2^{k-p} designs may be generated by selecting generators in such a way that resolution III design gives some clear main effects along with some clear two-factor interaction while resolution IV design gives all clear main effects without any clear two-factor interaction. If the objective is to investigate some main effects along with some two-factor interaction rather than all main effects then the resolution III design will be the best choice compared to the resolution IV design. Maximum resolution criteria in this case will give the wrong recommendation as stated in Ref. [2].
[b]Among many resolution IV designs with a given factor k and $1/2^p$th fraction, the designs having largest number of clear two-factor interactions are best. For this case minimum aberration criteria do not give a better design [20].
[c]Three-factor interaction usually remains insignificant.

6.6 REGRESSION MODELING

The results of a designed experiment can be used to develop an empirical model that relates the input variables with the response. Such a model is useful for prediction and control. This modeling is usually done by regression analysis. This model also helps in selecting conditions that lead to optimum response. The polynomial model normally used is:

$$y = b_0 + \sum_{i=1}^{k} b_i x_i + \varepsilon \qquad (6.19)$$

where

y = response of the experiment,
b_i = coefficients found by regression, $0 \leq i \leq k$,
k = number of regressor variables,
x_i = regressor representing *linear* (e.g., main factors) or *nonlinear* variables (e.g., interaction and quadratic terms of main factors),
ε = error or residual.

Although in the above model variables may be linear or nonlinear, the model is linear in coefficients b_i and is called a *linear regression model*. The coefficients are found by the linear least square method, which minimizes the error ε. In order to develop the method to find the coefficients let us express the model in terms of all experimental observations in matrix form as

$$Y = Xb + \varepsilon \qquad (6.20)$$

where

Y = Column matrix of N observed responses having order $N \times 1 = [y_1 \quad y_2 \ldots y_N]^T$.
X = Regressor variable matrix of order $N \times p$, where $p = k + 1$ represents the total number of coefficients in the model including the constant

$$= \begin{bmatrix} 1 & x_{11} & x_{12} & \cdots & x_{1k} \\ 1 & x_{21} & x_{22} & \cdots & x_{2k} \\ \vdots & & & & \\ 1 & x_{N1} & x_{N2} & \cdots & x_{Nk} \end{bmatrix}$$

where x_{ij} represents level of jth regressor variable at ith experiment.

Note that if the levels are expressed in coded form in the model the relative importance of the factor can be easily deduced.

b = coefficient matrix of order $p \times 1 = \begin{bmatrix} b_0 & b_1 & \ldots & b_k \end{bmatrix}^{\mathrm{T}}$,

ε = column matrix of errors or residuals at N experiments of order

$$N \times 1 = \begin{bmatrix} \varepsilon_1 & \varepsilon_2 & \ldots & \varepsilon_N \end{bmatrix}^{\mathrm{T}}$$

The criteria used in the least square method to find the coefficients is to minimize the deviation of the predicted value from the experimental value of response. This is achieved by minimizing the following function:

$$f = \sum_{i=1}^{N} \varepsilon_i^2 = (Y - Xb)^{\mathrm{T}}(Y - Xb) \tag{6.21}$$

Using matrix operations the above equation can be expressed as

$$f = Y^{\mathrm{T}}Y - 2b^{\mathrm{T}}X^{\mathrm{T}}Y + b^{\mathrm{T}}X^{\mathrm{T}}Xb \tag{6.22}$$

The minimization principle (i.e., equating to zero the partial differential of function f with respect to unknown coefficients) will give the relationship $-2X^{\mathrm{T}}Y + 2X^{\mathrm{T}}Xb = 0$. Rearranging it gives the coefficient matrix as

$$b = (X^{\mathrm{T}}X)^{-1}X^{\mathrm{T}}Y \tag{6.23}$$

The above equation can also be used in the model expressed as

$$y = b_0 x_1^{b_1} x_2^{b_2} x_3^{b_3} \tag{6.24}$$

where x_1, x_2, and x_3 are main factors. Taking the log of both sides the model becomes

$$\log_e y = \log_e b_0 + b_1 \log_e x_1 + b_2 \log_e x_2 + b_3 \log_e x_3 \tag{6.25}$$

If we take $\log_e y$ as response then this equation is linear in coefficients $\log_e b_0$, b_1, b_2, and b_3; hence Eqns (6.24) and (6.25) represent the linear regression model. Therefore the linear least square equation (6.23) can be used to estimate coefficients of the above model.

In some engineering applications the model may be nonlinear in the coefficients. One form is given as

$$y = b_0(1 - x_1^{b_1}) + b_2 x_2 \tag{6.26}$$

The coefficients in the above equations are determined by using the nonlinear least square method. One such method developed by authors [12] has been used in characterization of wear, discussed in the next chapter.

6.6.1 Significance of the Model

Significance of the model is tested by doing variance analysis. The ANOVA table is given in Table 6.6. The nomenclature of symbols used are given after the table.

It should be noted that in Table 6.6, SS_T is partitioned into SS_{reg} and SS_{res} while SS_{res} is further partitioned into SS_{lof} and SS_{pe}. This information can be expressed by the following two equations:

$$SS_T = SS_{reg} + SS_{res} \tag{6.27}$$

$$SS_{res} = SS_{lof} + SS_{pe} \tag{6.28}$$

The model is tested in two ways:

1. The significance of regression is tested with reference to residual (i.e., MS_{res}). If F-ratio is found to be greater than $F_{\alpha,k,N-k-1}$ the model is said to be significant, which means that out of k regressor variables at least one regressor is significant.

Table 6.6 ANOVA table for regression model

Source	Sum of squares	Degree of freedom (df)	Mean squares	F-ratio
Regression	SS_{reg}	k	$MS_{reg} = SS_{reg}/k$	$\dfrac{MS_{reg}}{MS_{res}}$
Residual	SS_{res}	$N-k-1$	$MS_{res} = SS_{res}/(N-k-1)$	
Lack of fit	SS_{lof}	$df_{lof} = df_{res} - df_{pe}$	$MS_{lof} = SS_{lof}/df_{lof}$	$\dfrac{MS_{lof}}{MS_{pe}}$
Pure error	SS_{pe}	$\sum_{i=1}^{n_p}(n_i - 1) = N - n_p$	$MS_{pe} = SS_{pe}/(N - n_p)$	
Total	SS_T	$N-1$		

where
SS_{reg} = Sum of squares (SS) due to regression model = $b^T X^T Y - g^2/N$; g is grand total of all experimental observations y; N is total number of observations; b, X, and Y have already been defined in Eqn (6.20).
SS_{res} = Residual SS representing sum of squared deviation between experimental value and predicted value f as defined in Eqn (6.21).
n_p, n_i, N = Number of experimental points, number of replication/s at ith experimental point, and total number of experiments, respectively.
SS_{pe} = SS due to pure error = $\sum_{i=1}^{n_p}\sum_{j=1}^{n_i}(y_{ij} - \bar{y}_i)^2$; y_{ij} is the observation at ith experimental point and jth replicate; $\bar{y}_i$ is the average of observations at ith experimental point.
SS_{lof} = SS for lack of fit representing weighted sum of squared deviation between expected response $\bar{y}_i$ at ith location and predicted value $\hat{y}_i$ at this location = $\sum_{i=1}^{n_p} n_i(\bar{y}_i - \hat{y}_i)^2$
$SS_T = Y^T Y - g^2/N$

2. Nonsignificance of lack of fit is tested with reference to pure error (i.e., MS_{pe}). If F-ratio is found to be less than $F_{\alpha,df_{lof},df_{pe}}$ then the model is linear. If lack of fit is found to be significant then the model should be discarded and another model should be tried out.

Individual coefficients are tested by computing the F-ratio using the formula $F_i = (b_i^2/C_{ii})/MS_{res}$ where C_{ii} is the diagonal element of $(X^T X)^{-1}$ corresponding to the ith regressor variable x_i. If F_i, the F-ratio of the ith coefficient of b_i corresponding to regressor variable x_i is found to be greater than $F_{\alpha,1,N-k-1}$ then that coefficient is said to be significant. In other words it can be said that variable x_i is significant. Note that this is not a confirmatory test since the value of the coefficient depends on the other regressor variables in the model. For details, refer to the DOE literature [21].

6.6.2 Testing the Adequacy of the Model

Adequacy of the model is checked by plotting residuals, computing externally studentized residual t_i of all experiments, quantifying and comparing R-squared R^2, adjusted R-squared R^2_{adj}, and measuring the predictive capability of the model $R^2_{prediction}$.

Residual plots are helpful in checking whether the residuals are normally distributed since the testing hypothesis is based on this assumption.

Additional information is required is to find outliers. The outliers are the experimental points at which residual $y_i - \hat{y}_i$ goes beyond the specified limit (e.g., $\pm 3\sqrt{MS_E}$). The best way to find the outlier is to compute externally studentized residual t_i at all experimental points and to check the outlier through the hypothesis testing described in Section 6.2 since the statistic t_i is a t-distributed random variable with $N-p-1$ degree of freedom. The formula to calculate t_i is given below

$$t_i = \frac{y_i - \hat{y}_i}{\sqrt{(1 - h_{ii})S_{(i)}^2}} \qquad (6.29)$$

where

h_{ii} = Diagonal element of hat matrix H of order $N \times N$ defined as $H = X(X^T X)^{-1} X^T$. It can be easily shown that the predicted response column matrix $\hat{Y}$ is easily obtained from $\hat{Y} = HY$.

$S_{(i)}^2$ = Estimate of variance of residual σ^2 by removing the ith observation

$$= \frac{(N-p)\mathrm{MS_{res}} - (y_i - \hat{y}_i)^2/(1 - h_{ii})}{N - p - 1}.$$

The outliers found should be critically examined. This may be due to a serious fault in experiments, measurements, or wrong assumptions in modeling.

How close the experimental observations are fitted with the model is quantified by the coefficient of determination R^2 and can be obtained from the following equation:

$$R^2 = 1 - \frac{\mathrm{SS_{res}}}{\mathrm{SS_T}} \qquad (6.30)$$

By adding higher order terms whether it is significant or not the R^2 value will increase but it does not mean that the model is good. For a good model having a large number of significant terms adjusted R-squared R^2_{adj} should be estimated. In this statistic an adjustment is done for the corresponding df of $\mathrm{SS_{res}}$ and $\mathrm{SS_T}$. The equation is given below.

$$R^2_{adj} = 1 - \frac{\mathrm{SS_{res}}/(N-p)}{\mathrm{SS_T}/(N-1)} = 1 - \frac{(N-1)}{(N-p)}(1 - R^2) \qquad (6.31)$$

where N and p represent the total number of experiments and the total number of coefficients in the model including the constant, respectively.

The large difference in R^2 and R^2_{adj} implies that nonsignificant terms have been added in the model.

The capability to predict a new observation by the model is quantified by $R^2_{prediction}$, which is obtained by this formula:

$$R^2_{prediction} = 1 - \frac{\mathrm{PRESS}}{\mathrm{SS_T}} \qquad (6.32)$$

where

$$\mathrm{PRESS} = \text{Prediction sum of squares} = \sum_{i=1}^{N} \left(\frac{y_i - \hat{y}_i}{1 - h_{ii}}\right)^2 \qquad (6.33)$$

There should not be much difference between R^2_{adj} and $R^2_{prediction}$. The above terms described to check the adequacy of the model are given to understand the concept and are only indicative. Refer to the literature [1,21] for more detailed analysis. The empirical model obtained must be verified by the experiments at other locations within the design space.

NOMENCLATURE

A, B, C	three different types of factors
A_i	number of generators having i factors
A_r	number of generators having r factors, where r is smallest integer in the elements of pattern vector W such that $A_r \geq 1$
AB	two-factor interaction term, interaction between linear component of A and linear component of B
AC, BC	two-factor interaction terms, defined similarly as AB
AB^2	two-factor interaction term, interaction between linear component of A and quadratic component of B
AC^2, BC^2	two-factor interaction terms, defined similarly as AB^2
ABC	three-factor interaction term, interaction between linear component of A, linear component of B, and linear component of C
ABC^2	three-factor interaction term, interaction between linear component of A, linear component of B, and quadratic component of C
AB^2C, AB^2C^2	three-factor interaction terms, defined similarly as ABC^2
b	coefficient matrix of order $p \times 1$ where $p = k + 1$
b_i	coefficients in the regression model, $0 \leq i \leq k$
C_{ii}	diagonal element of $(X^T X)^{-1}$ corresponding to ith regressor variable x_i
d	nondimensional difference between means of true and hypothesized distribution defined as $\Delta / \sqrt{\sigma_1{}^2 + \sigma_2{}^2}$
d_i	$= y_{1i} - y_{2i}$
d_1, d_2	two different types of 2^{k-p} designs such that A_r of $d_1 \neq A_r$ of d_2
df	degree of freedom
$df_{lof}, df_{res}, df_{pe}$	degree of freedom for lack of fit, residual, and pure error
$E(y)$	expected value of y where E is called expected value operator
f	function giving sum of squares of errors or residuals
$f(y)$	probability density function of y if it is continuous
F_i	$= \dfrac{b_i{}^2 / C_{ii}}{MS_{res}}$, F-ratio for ith coefficient of b_i corresponding to regressor variable x_i
F_{n_1-1, n_2-1}	F-distribution with numerator df of $n_1 - 1$ and denominator df of $n_2 - 1$
F_0	test statistic used to compare the variances of two populations
F_{α, ν_1, ν_2}	upper percentile of the F-distribution at significance level α, with numerator df of ν_1 and denominator df of ν_2
g	grand total of all experimental observations
h_{ii}	diagonal element of hat matrix H
H	$X(X^T X)^{-1} X^T$, hat matrix of order $N \times N$
H_0	null hypothesis, the statement to be tested
H_1	alternative hypothesis, the condition to be concluded at rejection
I	generator to construct two-level fractional design
I	generator to construct three-level fractional design
$I_{smallest}$	the generator having the smallest number of factors
k	number of factors
k	number of regressor variables

l_1, l_2 number of levels in factors 1 and 2

$MS_{interaction}, MS_E$ mean squares due to interaction between factors and experimental error

$MS_{lof} MS_{pe}$ mean squares due to lack of fit and pure error

$MS_{reg} \; MS_{res}$ mean squares due to regression and residual

MS_1, MS_2 mean squares due to factors 1 and 2, mean squares are obtained by dividing SS with corresponding df

n sample size

n number of engine tests

n number of replications

n_i number of replication/s at ith experimental point

n_p number of experimental points

n_1, n_2 sample size of two random samples from populations 1 and 2

$n(I_{smallest})$ number of factors in $I_{smallest}$, called resolution of the design expressed in roman number

N total number of experiments

$N(\mu, \sigma^2)$ normal distribution with mean μ and variance σ^2

p $k + 1$, represents total number of coefficients in the model including constant

p number of independent generators to construct a fractional design

$p(y)$ probability function of y if it is discrete

PRESS   predicted sum of squares

r r is the smallest integer in the elements of W such that $A_r \geq 1$

R^2 coefficient of determination, R-squared

R^2_{adj} adjusted R-squared

$R^2_{prediction}$ predictive capability of the model

s_{m3} remainder of Sum/3, called modulus 3 (mod 3)

S sample standard deviation

S^2 sample variance

$S^2_{(i)}$ estimate of variance of residual σ^2 by removing the ith observation

S^2_1, S^2_2 sample variances of two random samples from populations 1 and 2

S_p square root of pooled sample variance

SS $= \sum_{i=1}^{n} (y_i - \bar{y})^2$, sum of squares

$SS_E, SS_{interaction}$ SS due to experimental error and interaction between factors

SS_{lof}, SS_{pe} SS due to lack of fit and pure error

SS_{reg}, SS_{res} SS due to regression model and residual representing sum of squared deviation between experimental value and predicted value

SS_T total sum of squares

SS_1, SS_2 SS of two random samples from populations 1 and 2

SS_1, SS_2 SS due to factors 1 and 2

$SS_{1+2+interaction}$ SS due to combined effect of factor 1, factor 2, and their interaction

t_i externally studentized residual at ith experimental point

$t_{n-1}, t_{n_1+n_2-2}$ t-distributions with $n-1$ df and $n_1 + n_2 - 2$ df, respectively

t_{01} test statistic used to compare the population mean μ with a fixed value μ_0 if σ is not known

t_{02} test statistic used to compare the means of two populations if σ is not known

$V(y)$	variance of y, where V is called variance operator
$\overline{w}$	$= W/(l_1 l_2 n)$, overall average wear rate
$\overline{w}_{f_1 i}$	average wear rate in ith row
$\overline{w}_{f_2 j}$	average wear rate in jth column
$\overline{w}_{ij}$	average wear rate at ijth cell
w_{ijk}	wear rate at ith level of factor 1, jth level of factor 2, and kth replication $1 \leq i \leq l_1$; $1 \leq j \leq l_2$; $1 \leq k \leq n$
W	$= (A_3, A_4, \ldots, A_k)$, pattern vector for 2^{k-p} design; that is, $1/2^p$th fraction of 2^k design
W	overall sum of wear rates
$W_{f_1 i}$	sum of wear rates in ith row
$W_{f_2 j}$	sum of wear rates in jth column
x_i	coded value of additive concentration at ith experiment
x_i	regressor representing a *linear* (e.g., main factors) or *nonlinear* variables (e.g., interaction and quadratic terms of main factors)
x_{max}, x_{min}	maximum and minimum values of additive concentrations in coded form
X	regressor variable matrix of order $N \times p$ where $p = k + 1$
y	random variable
y	response of the experiment
y_i	element of a sample of size n
y_{ij}	experimental observation at ith experimental point and jth replicate
$\overline{y}$	$= \sum_{i=1}^{n} y_i \Big/ n$, sample mean
$\overline{y}_i$	average of observations at ith experimental point
$\hat{y}_i$	predicted value at ith experimental point
$\overline{y}_1, \overline{y}_2$	sample means of two random samples from populations 1 and 2
y_1, y_2	two random variables
y_{1i}, y_{2i}	wear coefficients due to lubricants 1 and 2 of ith specimen
Y	column matrix of N observed responses having order $N \times 1$
$\hat{Y}$	HY, predicted response column matrix of order $N \times 1$
z	standard normal random variable defined as $(y-\mu)/\sigma$
z_i	additive concentration in % (w/v) at ith experiment
z_{max}, z_{min}	maximum and minimum values of additive concentrations in % (w/v)
z_0	additive concentration in % (w/v) at mid level (i.e., at the center point)
z_{01}	test statistic used to compare the population mean μ with a fixed value μ_0 if σ is known
z_{02}	test statistic used to compare the means of two populations if σ is known
$z_\alpha, z_\beta, z_{\alpha/2}$	standard normal random variables at significance level α, β, and $\alpha/2$, such that the area under the standard normal distribution curve from z_α to ∞ is α, from z_β to ∞ is β, and from $z_{\alpha/2}$ to ∞ is $\alpha/2$

Greek Letters

α	level of significance, which is the probability of Type-I error
β	probability of Type-II error
χ^2_{n-1}	Chi-square distribution with $n - 1$ df
χ^2_0	a test statistic used to compare the population variance σ^2 with a fixed value σ^2_0

Δ the difference between means of true and hypothesized distribution of $\bar{y}_1 - \bar{y}_2$

Δz step value, $= (z_{max} - z_{min})/(x_{max} - x_{min})$, or $= (\log_e z_{max} - \log_e z_{min})/(x_{max} - x_{min})$ depending on the model

ε error or residual

ε column matrix of errors or residuals at N experiments of order $N \times 1$

ε_i residual, the difference between experimental and predicted values at ith experiment, $1 \le i \le N$

μ population mean

μ_d mean of differences between wear coefficients due to lubricants 1 and 2 in a blocking experiment

μ_0 fixed mean value

μ_1, μ_2 population mean of random variables y_1 and y_2

σ standard deviation

σ^2 population variance

σ^2 variance of residual

σ_0^2 fixed variance value

σ_1^2, σ_2^2 population variance of random variables y_1 and y_2

REFERENCES

[1] Montgomery DC. Design and analysis of experiments. 5th ed. New York, NY: John Wiley and Sons; 2003.

[2] Wu CFJ, Hamada M. Experiments—planning, analysis, and parameter design optimization. New York, NY: John Wiley and Sons; 2002.

[3] Fisher RA. The design of experiments. 4th ed. Edinburgh: Oliver and Boyd; 1947.

[4] Box GEP, Wilson KB. On the experimental attainment of optimum conditions. J R Stat Soc B 1951;13(1):1—45.

[5] Nair VN, editor. Taguchi's parameter design: a panel discussion. Technometrics 1992;34(2):127—61.

[6] Meyer PL. Introductory probability and statistical applications. 2nd ed. Reading, MA: Addison-Wesley; 1970.

[7] Hogg RV, Craig AT. Introduction to mathematical statistics. 5th ed. Upper Saddle River, NJ: Pearson Education; 1995.

[8] NIST/SEMATECH e-Handbook of Statistical Methods, <http://www.itl.nist.gov/div898/handbook/>.

[9] NORMSDIST() function in statistical category of MS-Excel.

[10] Montgomery DC, Runger GC. Applied statistics and probability for engineers. 3rd ed. New York, NY: John Wiley and Sons; 2003.

[11] Kumar R, Prakash B, Sethuramiah A. A methodology to estimate the wear coefficient of engine liner under lubricated condition. Indian J Trib 2008;3(1):15—19.

[12] Kumar R, Prakash B, Sethuramiah A. A systematic methodology to characterise the running-in and steady state wear process. Wear 2002;252(5—6):445—53.

[13] Mason RL, Gunst RF, Hess JL. Statistical design and analysis of experiments-with applications to engineering and science. 2nd ed. New York, NY: John Wiley and Sons; 2003.

[14] Neema ML, Pandey PC. Investigation of the performance characteristics of cold-worked machined surfaces. Wear 1980;60(1):157—66.

[15] Choudhury IA, El-Baradie MA. Tool-life prediction model by design of experiments for turning high strength steel (290 BHN). J Mat Proc Tech 1998;77: 319−26.

[16] Hamada M, Wu CFJ. Analysis of designed experiments with complex aliasing, IIQP research report, RR-91-01; 1991.

[17] Box GEP, Hunter JS. The 2^{k-p} fractional factorial designs part I. Technometrics 1961;3(3):311−51.

[18] Box GEP, Hunter JS. The 2^{k-p} fractional factorial designs part II. Technometrics 1961;3(4):449−58.

[19] Fries A, Hunter WG. Minimum aberration 2^{k-p} designs. Technometrics 1980;22(4): 601−8.

[20] Chen J, Sun DX, Wu CFJ. A catalogue of two-level and three-level fractional factorial designs with small runs. Int Stat Rev 1993;61(1):131−45.

[21] Draper NR, Smith H. Applied regression analysis. 3rd ed. New York, NY: John Wiley and Sons; 1998.

CHAPTER 7

Detailed Methodology for Chemical Wear Modeling

7.1 INTRODUCTION

The modeling of chemical wear is of significant importance. As discussed in Chapter 5, the present state of knowledge is inadequate to theoretically predict the wear behavior.

Wear characterization at present is based mostly on dimensional change over a specific period of running time. Wear comparisons are done simply on the basis of wear scar dimension. In some cases the comparisons may be based on calculated wear volume or weight loss. The wear behavior for different additives, for example, is based on scar diameters obtained in a 4-Ball machine. The test machines can be standard or in-house. The standard machines have undergone a lot of up-gradation and can be considered excellent in design. However with regard to wear characterization, the approaches are inadequate despite the knowledge base available. It is well known that the wear processes involve running-in and steady state. Any meaningful comparisons should take this into account. The second aspect is how to effectively model the wear process mathematically.

The first section demonstrates the importance of separating running-in and steady-state wear with a typical example. The next section considers a known empirical approach to wear modeling and its application to wear tests conducted with engine oil in a reciprocating tester over long duration. Designed experiments have been conducted and the wear behavior in running-in and steady state are characterized by empirical equations. The mathematics involved is thoroughly discussed and the computer program used has been annexed at the end of the book. The importance of this approach has been emphasized in these sections. The final section addresses the practical problem of shorter test durations and how best to compare the wear behavior in such cases with an example.

It may be noted that the wear models in this chapter are based on a boundary lubrication regime. Depending on test geometry partial hydrodynamic/EHL effects may occur particularly in long duration tests. Such

Modeling of Chemical Wear.
DOI: http://dx.doi.org/10.1016/B978-0-12-804533-6.00007-X
© 2016 Elsevier Inc.
All rights reserved.

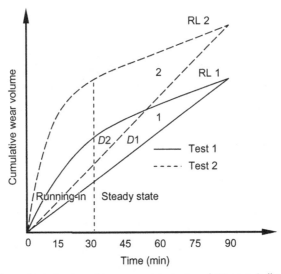

Figure 7.1 Schematic behavior of two repeat tests 1 and 2 in a 4-Ball tester. RL refers to regression line and D refers to the line drawn from the origin to the final wear volume.

effects if any are easily discernible from falling friction coefficient curve. In such cases the wear study may be confined to the boundary friction zone.

7.2 WEAR CHARACTERIZATION

Wear characterization usually done by dimensional change such as measuring wear scar diameter in a 4-Ball machine is not adequate to distinguish between effects of different variables such as additives and operating conditions on wear. This has been discussed in Section 4.4.2 of Chapter 4 with an example. In that section it was emphasized that wear should be evaluated progressively in terms of wear volume so that wear rates during running-in and steady state conditions may be separated. The running-in is the initial period of wear during which gradual adjustment of surfaces takes place. In this period the wear rate is high. After a certain period called the running-in period the wear rate stabilizes to a lower value. The running-in and steady-state wear behavior of two repeat tests 1 and 2 in a 4-Ball machine has been shown schematically in Figure 7.1.

The plots are made on the basis of cumulative wear volume as a function of time with 15 min intervals. The lower ball is not disturbed and at each stage the wear volume is obtained by the average wear scar diameter and the operation continued until the end of the test. The average wear scar diameter can be estimated as per ASTM D 4172 in which the arithmetic average of six wear scar measurements on three lower balls is taken. On each lower ball two wear scar diameter measurements in the direction of sliding and perpendicular to it are taken. Mean wear scar diameter as per ASTM D 4172 in equation form is given below.

$$\text{Mean wear scar diameter, } d = \frac{ds_1 + dp_1 + ds_2 + dp_2 + ds_3 + dp_3}{6} \quad (7.1)$$

where ds_i and dp_i are the wear scar diameters in the direction of sliding and perpendicular to it, respectively, in ith ball.

The present authors consider improvement in the above estimation method is necessary. Since the wear scar found is elliptical, we have to obtain an equivalent circular diameter d from the summation of the elliptical scar area on the three lower stationary balls. The formula is:

$$3\frac{\pi d^2}{4} = \frac{\pi ds_1 dp_1}{4} + \frac{\pi ds_2 dp_2}{4} + \frac{\pi ds_3 dp_3}{4}$$

Hence the improved mean wear scar diameter,

$$d = \sqrt{\frac{ds_1 dp_1 + ds_2 dp_2 + ds_3 dp_3}{3}} \quad (7.2)$$

The difference in the mean wear scar diameter using these two methods increases if the ratio between the major and minor axes of elliptical wear scar increases. The large difference in wear scar estimation will result in a wrong estimation of wear coefficients if Eqn (7.1) is used.

After estimating mean wear scar diameter d from Eqn (7.2) the wear volume at different intervals can be obtained by using Eqn (4.10), which takes elastic recovery into account. The wear volume so obtained should be used for obtaining wear rates. Note that Figure 7.1 is only a schematic diagram of cumulative wear volume versus time, and the preceding discussion refers to the right approach to get the cumulative volume. The wear rate in terms of wear volume per unit time considering only the final wear volumes are obtained from the slopes of the $D1$ and $D2$ lines. It is evident that significant variation can occur on this basis. If it is assumed that running-in is completed in 30 min, the slope of the regressed line between 30 and 90 min gives the steady-state wear rate. The regressed lines are shown as RL1 and RL2 for the two tests. For clarity the regressed line is shown as a part of the overall curve. The wear rates will be lower now as compared to the case based only on the final wear volume. Better repeatability is also expected by such a procedure as the variations due to running-in effects are reduced. The wear rates need not necessarily represent steady wear rates but the above example clearly brings out the importance of defining the wear rate.

This procedure needs to be assessed from a basic point of view by comparison with detailed modeling based on long duration tests presented in Section 7.3.1. A detailed methodology to determine wear rate as a function of operating variables is presented in Section 7.3.2. This will be very useful in mapping wear behavior in a test machine. Programs developed on this basis will be useful to the lubricant formulators and users for developing wear maps. One additive can be better or worse than the other depending on the operating conditions and the nature of the additive. Distinction based on one or two conditions is inadequate for a formulator to make an effective judgment. Detailed investigations are possible through cooperative effort. One interesting example of such an effort was the study of transitions in lubrication regimes under different operating conditions by the International Research Group (IRG) on wear [1]. Similar approaches in wear characterization will be very helpful.

Long duration tests are time consuming and usually short duration tests are conducted for assessment. The knowledge base available in Section 7.3 with long duration tests can be appropriately tailored for this purpose albeit with some limitations. A detailed consideration of short duration tests is given in Section 7.4.

7.3 DETAILED EMPIRICAL WEAR MODELING

The fundamental approaches to chemical wear modeling have been discussed in Chapter 5. These models are inadequate to predict wear and hence cannot be used for developing wear maps. The solution lies in modeling wear empirically. The empirical wear modeling demands accurate and precise estimation of wear to find whether change in operating conditions and lubricant affect wear rates, particularly when changes are small. Hence the methodology to quantify the running-in and steady-state wear rates must be based on a proper mathematical model and statistical procedures. The proposed model adopts the mathematical formulation of the running-in and steady-state wear by Zheng et al. [2]. The coefficients in the mathematical expression are of nonlinear nature.

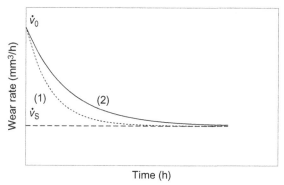

Figure 7.2 Wear rate variation with time.

A new methodology has been developed to estimate the coefficients. Methodology for the running-in period has also been developed in the proposed model described in Subsection 7.3.1. The wear rate data so obtained were used in developing an empirical wear model as a function of operating conditions. This is based on experimental design and discussed in detail in Subsection 7.3.2.

7.3.1 Mathematical Model for Wear Rate

The present wear model characterizes running-in wear rate at time zero, running-in period and steady-state wear rate from the corresponding experimental wear data. The typical variation of wear rate of a new component with time has been shown schematically in Figure 7.2.

Initially the wear rate is high, which reduces exponentially with time and reaches steady-state after some time. This information may be mathematically expressed as follows [2]:

$$\dot{v} = (\dot{v}_0 - \dot{v}_s)e^{-bt} + \dot{v}_s \tag{7.3}$$

In this equation, coefficient b has an inverse relation with the running-in period. The influence of b is illustrated in Figure 7.2. Curve 1 corresponds to a higher b value in comparison to curve 2. In the case of curve 1 the running-in is faster. It should be noted that curves 1 and 2 shown in Figure 7.2 demonstrate the influence of b while running-in wear rate and steady-state wear rate were kept at the same value in both cases. In reality both values can vary depending on the operating conditions.

Putting $t = 0$ and $t = \infty$ in Eqn (7.3) we get $\dot{v}_0$ and $\dot{v}_s$, respectively, which satisfy boundary conditions of the wear process. Integrating Eqn (7.3) with respect to time t, we get

$$v = - \left(\frac{\dot{v}_0 - \dot{v}_s}{b} \right) e^{-bt} + \dot{v}_s t + a \tag{7.4}$$

At $t = 0$ wear volume $v = 0$. Putting this condition in Eqn (7.4) we get

$$a = \left(\frac{\dot{v}_0 - \dot{v}_s}{b} \right) \tag{7.5}$$

Combining Eqns (7.4) and (7.5) we find

$$v = \left(\frac{\dot{v}_0 - \dot{v}_s}{b} \right)(1 - e^{-bt}) + \dot{v}_s t \tag{7.6}$$

or,

$$v = a(1 - e^{-bt}) + \dot{v}_s t \tag{7.7}$$

For steady-state at $t = \infty$,

$$v = a + \dot{v}_s t \tag{7.8}$$

In Eqn (7.7), v is nonlinear in terms of coefficients a, b, and $\dot{v}_s$. This equation was linearized by assuming the value of b and then calculating a and $\dot{v}_s$ from the least square method given by following equation, discussed in Section 6.6:

$$A = (X^T X)^{-1} X^T Y \tag{7.9}$$

The X, Y, and A matrices are

$$X = \begin{bmatrix} 1 - e^{-bt_1} & t_1 \\ 1 - e^{-bt_2} & t_2 \\ \vdots & \vdots \\ 1 - e^{-bt_m} & t_m \end{bmatrix}; \quad Y = \begin{bmatrix} v_1 \\ v_2 \\ \vdots \\ v_m \end{bmatrix}, \quad \text{and} \quad A = \begin{bmatrix} a \\ \dot{v}_s \end{bmatrix} \tag{7.10}$$

where m = number of wear volume measurement during the wear test.

The calculation of nonlinear coefficients a, b, and $\dot{v}_s$ in Eqn (7.7) is programmed in MATLAB and is annexed and includes an explanation on use. The initial part of the program arrives at two values b_1 and b_2 iteratively. The value of b_1 corresponds to the condition where $(1-R)$ is negative and b_2 corresponds to the condition where $(1-R)$ is positive. The value of R is calculated by the following formula, which can easily be obtained from the basic definition of R-squared, R^2 given in Section 6.6.1.

$$R = \sqrt{\frac{\sum_{i=1}^{m}(\hat{v}_i - \bar{v})^2}{\sum_{i=1}^{m}(v_i - \bar{v})^2}} \qquad (7.11)$$

where

v_i = wear volume at time t_i (mm³),

$\bar{v}$ = average of all v_i,

$\hat{v}_i$ = wear volume at time t_i from predictor Eqn (7.6) (mm³).

Then bisection method is used to find the value of b iteratively after fixing the criteria that $|(1-R)| < \varepsilon_1$. ε_1 is a very small value and selected as 1×10^{-4} so that the coefficient of determination R approaches unity. This methodology finally gives the value of b, running-in wear rate $\dot{v}_0$, and steady-state wear rate $\dot{v}_s$. It may be clarified that running-in wear rate refers to the wear rate at the start of the test.

The running-in period t_r is calculated by defining it as the time at which a percentage of the wear rate equals steady-state wear. This may be expressed mathematically by the following equation:

$$t_r = -\frac{\log_e\left[\frac{\dot{v}_s}{(\dot{v}_0 - \dot{v}_s)}\left(\frac{100}{p} - 1\right)\right]}{b} \qquad (7.12)$$

where p is the selected percentage.

7.3.2 Empirical Wear Modeling Using Design of Experiments

Empirical wear modeling can be done using DOE. Authors have conducted a laboratory experiment on the SRV Optimol reciprocating tester under lubricating conditions with a ball-on-flat geometry. The material was En 31 steel and the lubricant used was engine oil [3]. The variables were load, temperature, and initial roughness. This geometry enables monitoring the small changes in wear volume by observing the change in scar diameter. This wear volume data at different time intervals with statistically designed experiments was used in assessing running-in wear rate, running-in period, and steady-state wear rate by using the MATLAB program described in the previous section. After wear characterization the empirical equations for running-in wear rate, running-in period, and steady-state wear rate were expressed as a function of the selected variables using the least square method. These equations may be used for developing wear maps. The details of designed experiments and empirical modeling are given below.

In the present study the experiments were performed by selecting one-third fraction of 3^3 fractional design given in Section 6.5.2. This design estimates the following four effects each having two df: $A + BC + ABC$, $B + AC^2 + ABC^2$, $C + AB^2 + AB^2C$, and $AB + AC + BC^2$ where factors A, B, and C are load (P), initial roughness (R_q), and contact temperature (T_c), respectively. In the first effect it is assumed that there is no interaction between roughness and temperature. It enables the influence of the first main factor load to be estimated reasonably upon considering two- and three-factor interactions are insignificant. Similarly the influence of the main factors roughness, and temperature are estimated. Hence main factors can be easily analyzed from the above design. The fourth effect will estimate combined effect of load—roughness, load—temperature, and roughness—temperature interactions if it exists. This three-level design enables observation of the nonlinear nature of response. One additional experiment was done at each design point to estimate the error of experiments. Thus a total number of 18 experiments were conducted with this design.

The experimentation was done by reciprocating the upper ball on the stationary lower specimen. Three different types of roughness, 0.35, 0.55, and 0.75 μm, of the lower specimen were taken, which were obtained by grinding with different depths of cut. A drop of commercial engine oil was applied at the contact zone for lubrication. The quantity applied was approximately 0.1 ml. The formulated oil contains zinc dithiophosphate plus detergent and dispersant additives. Zinc and phosphorous were 742 and 1890 ppm in the oil. Sulfur content was not measured. The viscosity of the lubricant was 129.86 cSt at 40 °C and 13.29 cSt at 100 °C. The ball was reciprocated perpendicular to the lay of roughness, selecting stroke length of 1 mm and frequency of 50 Hz. The load was varied as 20, 40, and 60 N while temperature was varied as 50, 100, and 150 °C. Total duration was 8 h with 10 steps of 10, 10, 20, 20, 30, 30, 60, 60, 120, and 120 min. At each stage the specimens were cleaned with benzene and acetone and the ball scar diameter was measured along and across the sliding direction. The geometric mean of these two values was taken for calculating the wear volume. The ball was relocated at each stage and a fresh drop of oil was added. Friction was continuously recorded throughout the test.

Ball wear volume was calculated using following equation, which takes elastic recovery into account [4]:

$$v = \frac{\pi d_0{}^4}{64r} \left[\left(\frac{d}{d_0}\right)^4 - \frac{d}{d_0} \right] \qquad (7.13)$$

where

$$d_0 = 2 \left(\frac{3Pr}{4E}\right)^{1/3} \qquad (7.14)$$

$$\frac{1}{r} = \frac{1}{r_1} \pm \frac{1}{r_2} \qquad (7.15)$$

$$\frac{1}{E} = \frac{1 - \nu_1^2}{E_1} + \frac{1 - \nu_2^2}{E_2} \qquad (7.16)$$

$$d = \sqrt{d_s d_p} \qquad (7.17)$$

where d_0 is Hertzian diameter, d_s and d_p are diameters of ball scar along and across the sliding direction. ν_1 and ν_2 are Poisson's ratios and E_1 and E_2 are elastic constants of materials 1 and 2, respectively. r_1 and r_2 are the radii of surfaces 1 and 2 in contact and P is the load.

The wear volume calculated for the ball at different times was used as input to find running-in wear rate, running-in period, and steady-state wear rate. One sample of input, output, and comparison between theoretical and experimental values obtained from the MATLAB program is given in Figure 7.3.

The surface temperature rise calculations were based on the geometric contact area as discussed in Section 3.3. Since progressive wear increases the geometric area of contact, the temperature rise at 10 min, at the time of completion of running-in, and at 8 h of test were calculated for each set of operating conditions. These temperatures are denoted as $\Delta\theta_{t_{0.17}}$, $\Delta\theta_{t_r}$, and $\Delta\theta_{t_8}$, respectively. To calculate $\Delta\theta_{t_r}$ the geometric area of contact was obtained after calculating the diameter of ball wear scar from Eqn (7.13) by numerical procedure. In this equation the wear volume v at the completion of the running-in period was obtained from Eqn (7.7). In the development of empirical relation for the running-in wear rate temperature rise was taken as zero since running-in wear rate is obtained at time $t = 0$. For developing empirical relations for the running-in period average of $\Delta\theta_{t_{0.17}}$

Input file: Experimental_Data.xls

Output file: result.txt

Time (t) in   Wear volume (V) in
   hours   mm^3
     0   0
  0.1667   1.66E−04
  0.3333   1.73E−04
  0.6667   2.14E−04
     1   2.25E−04
    1.5   2.36E−04
     2   2.64E−04
     3   2.95E−04
     4   3.11E−04
     6   4.10E−04
     8   5.18E−04

(a)

```
running-in wear rate(Vo)=
1.676441e-003
steady state wear
rate(Vs)= 3.986185e-005
running in period(Tr) =
0.729795
a = 1.793511e-004
b = 9.125000
```

(b)

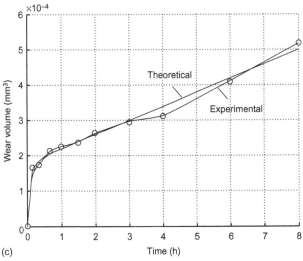

(c)

Figure 7.3 Sample of input (a), output (b), and comparison of experimental wear volume with the theoretical (regressed) curve obtained at 60 N load, 0.75 μm roughness, and 150 °C temperature from the MATLAB program (c).

and $\Delta\theta_{t_r}$ were taken as temperature rise, which represents the running-in zone. For developing the steady-state wear rate equation average of $\Delta\theta_{t_r}$ and $\Delta\theta_{t_8}$ was taken as temperature rise. The complete results at different operating conditions for steady-state wear rate along with surface temperature rise during this period are given in Table 7.1.

Variance analysis of steady-state wear rate having the main factors of load (P), initial roughness (R_q), and contact temperature (T_c), was carried out using the equations given in Section 6.4.2 and summarized in

Table 7.1 Steady-state wear rate and average surface temperature rise between running-in period and end of test at different operating conditions in ball-on-flat test

Serial number of experiment	Order of experiment	Load P (N)	rms value of roughness R_q (μm)	Bulk temperature T_b (°C)	Replication number	Steady-state wear rate w_s ($\times 10^{-5}$ mm^3/h)	Average surface temperature rise between running-in period and end of test $\Delta\theta$ (°C)
1	12	20	0.35	150	1	1.00	6
2	13	20	0.35	150	2	1.59	5
3	15	40	0.35	100	1	1.15	9
4	3	40	0.35	100	2	1.97	8
5	2	60	0.35	50	1	4.00	7
6	7	60	0.35	50	2	3.44	8
7	9	20	0.55	50	1	2.19	4
8	16	20	0.55	50	2	2.60	4
9	17	40	0.55	150	1	1.34	9
10	10	40	0.55	150	2	1.54	7
11	1	60	0.55	100	1	3.40	9
12	18	60	0.55	100	2	4.15	9
13	5	20	0.75	100	1	1.38	5
14	4	20	0.75	100	2	2.72	5
15	6	40	0.75	50	1	8.69	6
16	14	40	0.75	50	2	9.24	5
17	8	60	0.75	150	1	3.99	8
18	11	60	0.75	150	2	4.53	8

Table 7.2. F-ratios in the first three effects—28, 50.33, and 46.33—were found to be large compared to F-tabulated of 4.26 at a significance level of 0.05. Assuming roughness and the temperature interaction term along with three order terms as insignificant as discussed earlier it can be concluded that main factors significantly affect steady-state wear rate. The roughness is found to be the most significant factor followed by temperature and load. However the fourth effect, $PR_q + PT_c + R_q T_c^2$, was also found to be significant. It shows that load–roughness, load–temperature, and roughness–temperature interactions together have a significant influence on steady-state wear rate. It needs detailed study by taking full-factorial or RSM design. In the present study the focus was only to study

Table 7.2 ANOVA table for steady-state wear rate having main factors: load (P), initial roughness (R_q), and contact temperature (T_c)

Source of variation	Sum of squares ($\times 10^{-9}$)	Degree of freedom (df)	Mean sum of squares ($\times 10^{-9}$)	F-ratio	$F_{0.05, \nu_1, \nu_2}$
$P + R_q T_c + PR_q T_c$	1.67	2	0.84	28.00	4.26
$R_q + PT_c^2 + PR_q T_c^2$	3.01	2	1.51	50.33	4.26
$T_c + PR_q^2 + PR_q^2 T_c$	2.77	2	1.39	46.33	4.26
$PR_q + PT_c + R_q T_c^2$	1.81	2	0.91	30.33	4.26
Error	0.23	9	0.03		

the influence of main factors, hence one-third fraction of 3^3 fractional design has been adopted.

To develop empirical relations for steady-state wear rate w_s as a function of load P, initial roughness R_q, and contact temperature T_c, the assumed polynomial is given below. The contact temperature T_c is the sum of bulk temperature T_b and surface temperature rise $\Delta\theta$:

$$\log_e Y = a_0 + a_1 \log_e P + a_2 \log_e R_q + a_3 \log_e T_c \qquad (7.18)$$

The coefficients of the equations were found by Eqn (7.9) where X, Y, and A matrices are

$$X = \begin{bmatrix} 1 & \log_e P_1 & \log_e R_{q_1} & \log_e T_{c_1} \\ \vdots & \vdots & \vdots & \vdots \\ 1 & \log_e P_n & \log_e R_{q_n} & \log_e T_{c_n} \end{bmatrix}; \quad Y = \begin{bmatrix} \log_e w_{s_1} \\ \vdots \\ \log_e w_{s_n} \end{bmatrix};$$

$$A = \begin{bmatrix} a_0 \\ a_1 \\ a_2 \\ a_3 \end{bmatrix}, \quad \text{and} \quad n = 18 \qquad (7.19)$$

The coefficients a_0, a_1, a_2, and a_3 obtained are -8.9188, 0.7175, 0.9671, and -0.7786, respectively. The final equation for steady-state wear rate of the ball is:

$$w_s = 1.34 \times 10^{-4} P^{0.72} R_q^{0.97} T_c^{-0.78} \qquad (7.20)$$

Disc wear was estimated after the total test duration of 8 h. The profile of the worn scar was traced at the middle portion along the sliding direction using a Talystep instrument. Wear volume was then assessed by a developed computer program taking into account the three-dimensional shape of the worn scar of the disc. The disc and ball wear volume at 8 h were compared at different operating conditions. The linear relation between the two showed a correlation coefficient of 0.9935. The regression equation obtained is [3]:

$$v_{disc} = 1.46 v_{ball} + 1.52 \times 10^{-4} \qquad (7.21)$$

From this information it is reasonable to assume that the influence of the various parameters on the ball wear rate will have similar trends for the disc wear also. The difference in absolute values may be related to the microstructure of the disc as compared to that of the ball.

The variance analysis of the model shown in Eqn (7.20) was done using the equations given in Section 6.6.1. The F-ratio of the model was found to be 15.25 and is clearly larger than the tabulated value of 3.34 at significance level of 0.05. It shows that the assumed model effectively represents the experimental data. The R-squared R^2 and adjusted R-squared R^2_{adj} of the model were found to be 0.77 and 0.72, respectively. These values can be improved by assuming another model by adding more terms in the model. The improvement of the model was not done in this case. The improvement of the model is exemplified by taking another example in the next chapter. The influence of the factors on steady-state wear rate in the present case is discussed below on the basis of Eqn (7.20).

Steady-state wear was clearly influenced by roughness and temperature as well as load. It is of interest to see that the exponent of temperature is negative. Thus wear rate decreases with temperature in this case. Steady-state wear rate also decreases as the initial roughness decreases. The strong influence of initial roughness was unexpected as the final scar roughness was nearly the same for all three roughness values. It is normally considered that initial roughness mainly affects the running-in part of wear only. In fact the investigation done with regard to roughness had a practical aim of assessing whether initial roughness influenced steady-state wear. This is of importance in engines with regard to the life of the liner. The empirical relation shows the system-specific wear behavior and the need for such modeling. None of the effects can be predicted by the existing theories and the response to wear is specific to the system.

The observed relationships can be represented graphically where the influence of parameters can be effectively visualized. Wear rate as a function of roughness and temperature at a load of 20 N is shown in Figure 7.4

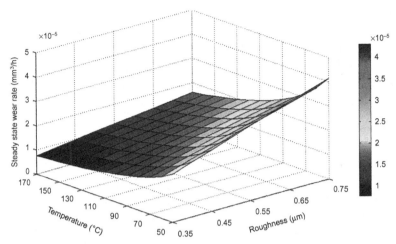

Figure 7.4 3D representation of steady-state wear rate as a function of roughness and temperature at 20 N load.

using MATLAB. The wear rates are given in mm^3/h. For comparison purposes the wear coefficient K described in Chapter 5 will be useful. The K values ranged from 7.79×10^{-9} to 3.85×10^{-8} in these experiments. These relatively low wear coefficients are typical of antiwear action. It may be observed that distinction at such low wear rates is effective through the adapted procedure. Limited analysis of films indicated their thickness ranged from 0.06 to 0.12 µm; that is again typical for this additive.

No detailed film analysis was carried out. The empirical relations can also lead to more realistic approaches to wear mechanisms. For example, the influence of temperature on steady–state wear rate may be reconciled with the possibility of more protective glassy films with higher nanohardness as temperature is increased. Recent evidence for such a possibility was discussed in Section 4.3.3 [5,6]. The conventional model would predict increasing wear with increasing temperature as discussed for the dry wear example in Section 5.3.2.4. In this case wear rate depended exponentially on the absolute temperature.

Another matter of importance is to decide when the running-in is complete. A possible practical criterion is to consider that running-in is complete when 95% of wear rate at a given point equals the steady-state rate. The running-in period was obtained on this basis and was related to the variables considered. Similarly the initial wear rate was also related to the variables involved. These relationships are not given here. It is of interest to note that the running-in time varied from 0.65 to 4.44 h depending on the operating conditions within a total run of 8 h. In repeat tests it was observed that the repeatability of initial wear rate is poorer in comparison to the steady-state wear rate. This may be attributed to the variability involved in the initial wear of the point contact. Thus arbitrary criteria used to determine steady-state wear in laboratory machines are unacceptable.

It may be argued that it is impractical to conduct long duration tests. But then we must keep in mind that significant errors are possible in short duration tests. Such a realization will help in ameliorating the present procedures, as we discuss in next section. It may be noted that in the present situation the running-in involves basically a changeover from point contact to area contact leading to steady state. In real systems running-in refers to micro-level adjustments. The present situation is treated as a running-in process in the sense that evolution of wear depends on the initial contact conditions that include load, temperature, and roughness. The methodology developed here is general and is applicable to wear study in any machine and test geometry. The empirical relationship can be used to construct wear maps.

7.4 PRACTICAL APPROACHES TO SHORT DURATION TESTS

Wear rate evaluation in short duration tests at the laboratory level is an important issue. Short duration tests are common in industry and it is necessary to understand how best to use such tests. This section is devoted to this issue and is based on the experiments conducted in a reciprocating rig and 4-Ball machines by the authors. There are many wear testers used in industry but the main approach to be used will be similar to those discussed here.

7.4.1 Reciprocating Rig

The wear tests conducted in the SRV Optimol reciprocating rig has been discussed in detail in a previous section. The wear volume was monitored intermittently at 10 points of a test of 8 h duration. The wear volume data are reprocessed using the MATLAB program discussed earlier by taking wear volume observations up to a total test duration of 6, 4, and 2 h. The coefficients a_0, a_1, a_2, and a_3 in empirical equation (7.18) obtained in each case are tabulated in Table 7.3 along with the values for 8 h of testing for comparison purposes. An influence of factors is similar for a 6-h test with a slight change in exponents. This order alters for 4 and 2 h. Further, the percentage change in steady-state wear rate values with

Table 7.3 Comparison of coefficients in empirical equations obtained for different time of test in reciprocating rig along with percentage change in wear rate with respect to 8 h of test

Total time of test (h)	Total number of wear measurements	Coefficients in empirical equation (7.18)				Percentage change in wear rate with respect to 8 h test
		Constant term a_0	Load exponent a_1	Roughness exponent a_2	Temperature exponent a_3	
8	10	-8.92	0.72	0.97	-0.78	
6	9	-8.72	0.68	0.89	-0.80	Range: 0–6, average: 3
4	8	-9.89	0.84	0.48	-0.73	Range: 2–26, average: 15
2	6	-7.39	0.53	0.29	-1.00	Range: 14–83, average: 41

respect to the values of 8 h at all experimental points were computed for 6, 4, and 2 h and is given in the last column of Table 7.3. The wear rate varies from 0% to 6% with an average of 3% for 6 h. On decrease of time the variation in wear rates increases significantly from 2% to 26% with an average of 15% for a 4-h test and from 14% to 83% with an average of 41% for a 2-h test. It may be concluded that if total time duration is lowered to 6 h from 8 h, the results will not be altered much with a maximum error of 6%. If time duration is further decreased the error increases rapidly. This may be due to a larger running-in period in the present case, which can extend up to 4.4 h. For the case of 4 and 2 h, influence of a running-in period causes erratic results. The large running-in period may be due to low contact stresses and initially a rougher surface of disc of 0.35–0.75 μm roughness. Hence the selection of total duration of the test depends on the starting point of the surface finish along with the operating conditions.

7.4.2 4-Ball Machine

The present subsection deals with relatively short duration tests conducted on a 4-Ball machine to distinguish two different formulations. Three possible approaches considered for evaluation are linear regression, nonlinear regression as done in the previous subsection, and nonlinear regression, assuming the slope at the final point to be the steady wear rate. The first two possibilities are discussed here and the last approach will be taken up in the next chapter.

The present tests were done for a total duration of 90 min with wear measurement after 15, 30, 60, and 90 min. Thus wear was characterized by a total of four points. Wear rate can be obtained by linear regression provided the point at which running-in is completed is known. For the present purpose the running-in was assumed to be completed within 30 min. Running-in time depends on the operating conditions. With a high level of loading and rotational speed as in the 4-Ball machine, the running-in duration may be short and 30 min duration may be acceptable. Thus wear was characterized by a total of four points. An example to determine wear rate based on regression of points between 30 and 90 min in a 4-Ball machine based on [7] is shown in Figure 7.5. The figure is based on the tabulated data given in Table 7.4. The tests were conducted at 40 kg and 1450 rpm. The purpose of the tests conducted was a systematic study of additive interactions. In this work three

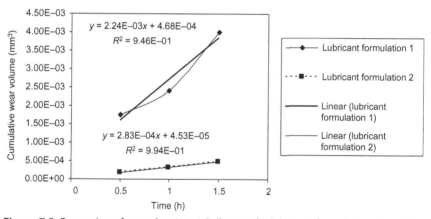

Figure 7.5 Regression of wear data in a 4-Ball tester for lubricant formulations 1 and 2.

Table 7.4 Wear data for two different lubricant formulations

Time (h)	Wear volume, mm^3 for lubricant formulation 1: ZDDP 0% (w/v), MoDTC 1% (w/v), and borate ester 1% (w/v)	Wear volume, mm^3 for lubricant formulation 2: ZDDP 2% (w/v), MoDTC 1% (w/v), and borate ester 1% (w/v)
0.25	8.44E−04	1.00E−04
0.5	1.75E−03	1.93E−04
1.0	2.40E−03	3.16E−04
1.5	3.99E−03	4.76E04

additives, namely ZDDP, MoDTC, and borate ester, were used to make different lubricant formulations in paraffinic base oil. Out of these, borate ester was commercially made whereas ZDDP and MoDTC were synthesized in the laboratory. Each formulation was stirred for 1 h using a magnetic stirrer to make a uniform suspension of additives in the base oil. The detailed optimization study has been discussed in the next chapter. Here only wear data of two tests in which ZDDP concentration is varied keeping MoDTC and borate ester concentration constant are given in Table 7.4.

The wear volume was obtained using Eqns (7.2) and (4.11) after measuring wear scar at different time intervals in the same test without dismantling the lower pot assembly. The wear scar diameter was measured by tilting the microscope eye piece at an angle around 28°. To get an exact perpendicular view of the scar this angle should be 35.3° from vertical, which can easily be found from the tetrahedral contact configuration of 4-Balls during the test. This angle could not be obtained due to the focusing limitation of the microscope. The oblique view of wear scar by 7° will not affect much in the wear scar diameter measurement and hence is accepted. The correlation coefficients in the above cases were found to be 0.973 and 0.997 for lubricant formulations 1 and 2, respectively. The wear rates were 2.24×10^{-3} and 2.83×10^{-4} mm^3/h for formulations 1 and 2, respectively, considering that running-in is completed in the initial 30 min as assumed. The concept of steady-state wear rate is based on equations in the earlier sections and mathematically amounts to infinite time and forms an effective base for comparison. The wear rates obtained by linear regression are just estimates of wear rates based on three points and may be called as linear wear rate.

The second method is to utilize the running-in equation and perform nonlinear regression as discussed in the previous section. This is based on the assumption that such regression with four points is only meaningful due to a short running-in period. Such regression assumes continuous increase in wear volume with time, which is normally the case. The regression equations obtained for two formulations by this procedure are

$$v = 6.83 \times 10^{-4}(1 - e^{-2.28t}) + 2.12 \times 10^{-3}t \qquad (7.22)$$

$$v = 5.33 \times 10^{-5}(1 - e^{-3.63t}) + 2.78 \times 10^{-4}t \qquad (7.23)$$

On this basis the results are tabulated in Table 7.5.

Table 7.5 Running-in wear rate, running-in period, and steady-state wear rates obtained by using annexed MATLAB program of nonlinear regression for two lubricant formulations

Parameters	For lubricant formulation 1	For lubricant formulation 2
Running-in wear rate	3.68×10^{-3} mm^3/h	4.71×10^{-4} mm^3/h
Running-in period	1.16 h	0.71 h
Steady-state wear rate	2.12×10^{-3} mm^3/h	2.78×10^{-4} mm^3/h

In both cases the correlation coefficients were >0.9999. The steady-state wear coefficients for formulations 1 and 2 were found to be 1.80×10^{-8} and 2.36×10^{-9}, respectively. This program takes all the data into account in estimating steady-state wear rate, running-in wear rate, and running-in period. This method can be used to classify wear rates accurately. It may be recalled that completion of the running-in is defined as the point at which 95% of the value equals steady-state wear rate. The running-in periods in the above cases were found to be 1.16 and 0.71 h and are lower than in the case of reciprocating rig as mentioned earlier. As previously discussed higher loads and speeds may be responsible for this.

This treatment based on steady-state wear worked well in the present example but was found to be inadequate in many other cases. This is mainly because 90 min is not sufficient to stabilize wear. This has been observed under many operating conditions discussed in the next chapter on optimization. Hence wear rate obtained from the slope at 90 min was used for comparison and was effective. This method will be elaborated in the next chapter along with the modified computer program.

It should be noted that simple linear regression does not ensure that the running-in influence is eliminated, though in the present example the wear rates are very close to the values obtained by nonlinear regression. *The best approach will be to use the nonlinear regression as a standard procedure wherever possible.* While issues of standardization depend on the agreement between the concerned organizations, the approaches may be adapted by researchers and formulators who are interested in distinguishing between different formulations more effectively.

A large number of wear tests are reported with 4-Ball machines at different conditions. The major interest in such studies is to distinguish between different additives used in formulations. While the nature of additives is getting sophisticated the approach to wear testing in many cases simply depends on the final wear scar diameter in a given test. The inadequacy of such an approach has been discussed earlier. Hein [8] has discussed the wear behavior on the basis of wear volume and wear coefficients. However the wear volume was that obtained at the end of the test. Another paper [9] used the concept of delta wear to characterize the wear behavior. Delta wear was defined as the difference between observed scar diameter and the Hertzian diameter at the selected load. The idea was perhaps to account for the elastic recovery. On the basis of Eqn (4.11) delta wear is not proportional to wear volume and it is better to calculate the wear volume directly by the equation. An interesting paper [10] considered step load tests but characterized the wear behavior on the basis of a step between 30 and 60 min only using the concept of delta wear as above. One more example of step load tests [11] is a comparison of wear scar with progressive loading up to 30 min with only 5 min steps. Influence of hexoxylborate was compared on this basis. It is obvious that present methods are ad hoc and there is an obvious need for a uniform procedure based on regression. While some investigations are reported with different operating conditions we have not come across any detailed wear mapping for antiwear additives.

The wear tests serve two purposes. One is the specification requirement imposed for Pass under defined test conditions. Such tests specify a maximum allowable scar diameter. The more important utility is in screening formulations. The confidence levels in screening are important and the first step is to have an effective procedure to clearly distinguish in a given machine. It is hoped that the suggested approaches including wear mapping will be useful for this purpose.

7.4.3 Wear in Real Systems

We need to emphasize that the ideas discussed so far are for amelioration of methodology in laboratory machines. This means that better comparison of wear rate as a function of operating conditions is possible and should be followed. However the comparison with real systems is not easy. The complexity of wear in real systems has been covered in Chapter 9 of LWST in detail and is not considered here. One of the major problems is to find the actual wear in real systems. Taking the example of the internal combustion engine the liner wear can be assessed by gauging provided adequate wear occurs. This usually translates to about 1000 h of running. Wear estimations with short duration tests are obviously beneficial. The authors have developed a methodology that can estimate wear coefficients in 200 h of endurance tests using a bearing area curve [12,13]. The wear volume was estimated from the bearing area curves before and after 200 h of tests. The method was also used to find the initial running-in wear over a period of 55 or 75 min as selected. The wear rate calculated in the endurance test was considered to be representative of the wear behavior. The wear rate in the endurance test, called steady-state wear, was found to be much lower than running-in wear as expected. The steady-state wear coefficients ranged from 2.52×10^{-10} to 6.84×10^{-10}. This example shows the need for better methods to estimate wear in real systems and is an on-going process.

NOMENCLATURE

a	integration constant
a_0, a_1, a_2, a_3	regression coefficients
A	regression coefficient column matrix
b	nonlinear coefficient having inverse relation with running-in period
d	mean wear scar diameter
d_p, d_s	diameters of ball scar across and along the sliding direction in a reciprocating rig
dp_i, ds_i	wear scar diameters in perpendicular and parallel to the direction of sliding in ith ball in a 4-Ball machine, $1 \leq i \leq 3$
d_0	Hertzian diameter
df	degree of freedom
D1, D2	lines drawn from origin to the final wear volume for tests 1 and 2
E	effective modulus of elasticity
E_1, E_2	elastic modulus of the two bodies in contact
$F_{0.05,\nu_1,\nu_2}$	upper percentile of the F-distribution at significance level 0.05, with numerator df of ν_1 and denominator df of ν_2
K	wear coefficient
m	number of wear volume measurement during the wear test
n	number of experiments
p	a percentage value, $p\%$, of wear rate at the running-in period equals steady-state wear rate
P	load
P_i	load at ith experiment, $1 \leq i \leq n$
PR_q	two-factor interaction term, interaction between linear component of P and linear component of R_q
PT_c, $R_q T_c$	two-factor interaction terms, defined similarly as PR_q
PR_q^2	two-factor interaction term, interaction between linear component of P and quadratic component of R_q
PT_c^2, $R_q T_c^2$	two-factor interaction terms, defined similarly as PR_q^2

$PR_q T_c^{\,2}$	three-factor interaction term, interaction between linear component of P, linear component of R_q, and quadratic component of T_c
$PR_q T_c$, $PR_q^{\,2} T_c$	three-factor interaction terms, defined similarly as $PR_q T_c^{\,2}$
r	relative radius of two surfaces at contact area before wear
r_1, r_2	radii of two surfaces 1 and 2 in contact
R	square root of coefficient of determination
RL1, RL2	regression lines for tests 1 and 2
R^2	coefficient of determination, R-squared
R_{adj}^2	adjusted R-square
R_q	initial rms roughness
R_{q_i}	initial rms roughness for ith experiment, $1 \le i \le n$
t	time
t_i	test time at ith wear volume measurement, $1 \le i \le m$
t_r	running-in period
T_b	bulk temperature
T_c	contact temperature
T_{c_i}	contact temperature at ith experiment, $1 \le i \le n$
v	ball wear volume
v	wear volume
v_i	wear volume at time t_i, $1 \le i \le m$
$\bar{v}$	average of all v_i, $1 \le i \le m$
$\dot{v}$	wear rate expressed as wear volume per unit time at time t
v_{ball}, v_{disc}	ball and disc wear volume after 8 h of test in reciprocating rig
$\hat{v}_i$	predicted wear volume at time t_i
$\dot{v}_0$, $\dot{v}_{t_r}$, $\dot{v}_s$	wear rate at time zero, running-in period, and steady state
w_s	steady-state wear rate
w_{s_i}	steady-state wear rate in ith experiment, $1 \le i \le n$
X	regressor variable matrix
Y	response column matrix

Greek Letters

$\Delta\theta$	surface temperature rise based on geometric contact area
$\Delta\theta_{t_{0.17}}$	surface temperature rise at 10 min of test
$\Delta\theta_{t_r}$	surface temperature rise at the time of completion of running-in
$\Delta\theta_{t_8}$	surface temperature rise at 8 h of test
ε_1	a very small value and selected as 1×10^{-4}
ν_1, ν_2	Poisson ratios of two bodies in contact

REFERENCES

[1] Salomon G. Failure criteria in thin film lubrication-the IRG programme. Wear 1976;36(1):1−6.
[2] Zheng M, Naeim AH, Walter B, John G. Break-in liner wear and piston ring assembly friction in a spark-ignited engine. Tribol Trans 1998;41(4):497−504.
[3] Kumar R, Prakash B, Sethuramiah A. A systematic methodology to characterise the running-in and steady-state wear processes. Wear 2002;252(5−6):445−53.

[4] Peterson MB, Winer WO. Wear control handbook. New York, NY: ASME; 1980. p. 449−50.

[5] Nicholls MA, Do T, Norton PR, Kasrai M, Bancroft GM. Review of the lubrication of metallic surfaces by zinc dialkyl-dithiophosphates. Trib Int 2005;38(1):15−39.

[6] Pereira G, Paniagua DM, Lachenwitzer A, Kasrai M, Norton PR, Capehart TW, et al. A variable temperature mechanical analysis of ZDDP-derived antiwear films formed on 52100 steel. Wear 2007;262(3−4):461−70.

[7] Saini H. A synergistic study and optimization of additive concentrations in engine oil using DOE [M. Tech. thesis]. Varanasi, India: Department of Mechanical Engineering, IIT(BHU); 2010.

[8] Hein RW. Evaluation of bismuth naphthenate as an EP additive. Lub Eng 2000;56(11):45−51.

[9] Weller Jr. DE, Perez JM. A study of the effect of chemical structure on friction and wear: Part I—Synthetic ester base fluids. Lub Eng 2000;56(11):39−44.

[10] Perez JM, Weller Jr. DE, Duda JL. Sequential Four-Ball study of some lubricating oils. Lub Eng 1999;55(9):28−32.

[11] Hu ZS, Dong JX, Chen GX. Synthesis and tribology of aluminium hexoxylborate as an antiwear additive in lubricating oil. Lub Sci 1999;12(1):79−88.

[12] Kumar R, Kumar S, Prakash B, Sethuramiah A. Assessment of engine liner wear from bearing area curves. Wear 2000;239:282−6.

[13] Kumar R, Prakash B, Sethuramiah A. A methodology to estimate the wear coefficient of engine liner under lubricated condition. Indian J Tribol 2008;3(1):15−19.

CHAPTER 8

Optimization Methodology in Additive Selection

8.1 INTRODUCTION

In formulating a lubricant, optimum concentration of additives is desired. This problem is multivariable and/or multiobjective. Selection of the best combination of additives to minimize or maximize single response involves a multivariable problem. When the problem involves optimization in terms of multiple responses like load carrying capacity as well as antiwear performance then the problem becomes multiobjective. Theoretical framework for multivariable optimization is first described followed by an example of work conducted in this area. In the next section, use of desirability function in solving a multiobjective problem is described followed by an example. Robust formulation of lubricant is also required so that there is the least possible variation in response due to a change of nuisance factors (the factors that cannot be controlled). This is discussed in the last paragraph of Section 8.3.2.3.

Optimization is a vast area and the mathematical coverage here is not exhaustive and only provides conceptual understanding. It is hoped that the examples given will clarify the practical approach to the problem and encourage use of the present and more advanced techniques in additive optimization. The emphasis in this chapter is on tribological performance.

8.2 MULTIVARIABLE OPTIMIZATION
8.2.1 Mathematical/Statistical Approach

Multivariable optimization can be defined as the methodology to find the value of variables at which the objective function value is either minimum or maximum. The objective function can be obtained empirically by regression modeling described in Chapter 6. The objective may be to minimize steady-state wear rate, coefficient of friction, or to maximize load carrying capacity. Let the objective function be expressed as $f(X) = f(x_1, x_2, \ldots, x_k)$ where vector $X = [x_1, x_2, \ldots, x_k]^T$ represents the

Modeling of Chemical Wear.
DOI: http://dx.doi.org/10.1016/B978-0-12-804533-6.00008-1
© 2016 Elsevier Inc.
All rights reserved.

k variables. These variables may be concentrations of different additives. A point $X^* = [x_1^*, x_2^*, \ldots, x_k^*]^T$ is called a stationary point if all the elements of $\nabla f = \left[\frac{\partial f}{\partial x_1} \frac{\partial f}{\partial x_2} \cdots \frac{\partial f}{\partial x_k} \right]^T$ are zero at X^*. This is called necessary condition of a stationary point. The stationary point will be minimum, maximum, or an inflexion point if the Hessian matrix $H = \nabla^2 f = \left[\frac{\partial^2 f}{\partial x_i \partial x_j} \right]$ of order $k \times k$ is positive definite, negative definite, or neither positive definite nor negative definite at X^*.

$$A \text{ matrix } H = \begin{bmatrix} H_{11} & H_{12} & \cdots & H_{1k} \\ H_{21} & H_{22} & \cdots & H_{2k} \\ \vdots & \vdots & \cdots & \vdots \\ H_{k1} & H_{k2} & \cdots & H_{kk} \end{bmatrix}$$

is called positive definite if and only if all the principal determinants

$$H_1 = |H_{11}|, \quad H_2 = \begin{vmatrix} H_{11} & H_{12} \\ H_{21} & H_{22} \end{vmatrix}, \ldots, \quad H_k = \begin{vmatrix} H_{11} & H_{12} & \cdots & H_{1k} \\ H_{21} & H_{22} & \cdots & H_{2k} \\ \vdots & \vdots & \cdots & \vdots \\ H_{k1} & H_{k2} & \cdots & H_{kk} \end{vmatrix}$$

are positive and it is negative definite if and only if the principal determinant H_j has sign $(-1)^j$ for $1 \leq j \leq k$. Positive or negative definiteness of a matrix can also be obtained from other methods mentioned in the optimization literature [1,2]. These conditions are called sufficient condition for an optimum point. If the optimum point lies within the experimental design space then it can be determined by simultaneously solving k equations obtained from necessary condition of an optimum point. Sometimes solution of these equations may not be easy if the objective function is complex nonlinear. In that case an appropriate iterative procedure given in literature may be adopted [2,3]. One versatile algorithm is a generalized reduced gradient (GRG) method if the objective function is differentiable. This algorithm may take care of constraints also. The constraints, for example, can be the following:

1. Additive concentrations should be within the prescribed limits of sulfated ash, phosphorous, and sulfur (SAPS) provided by euro norms.
2. Load carrying capacity should be greater than the prescribed value.
3. Wear coefficients should be less than the prescribed value.

The optimum solution can also be obtained by using the solver tool of Microsoft (MS) Excel. It is exemplified in Section 8.2.2.4. Other optimization software programs are MATLAB and Mathematica.

It should be noted here that if the optimum point lies within the experimental design space then the regression model must be of order ≥ 2 to express the curvature within the vicinity of optimum point. This model can be obtained either from the three-level factorial design discussed in Chapter 6 or central composite design (CCD), to be discussed in the last paragraph of this section.

If the optimum point does not lie within the region of experimentation the response surface methodology (RSM) developed by Box and Wilson [4] may be used to reach in the optimum region systematically and economically. In this methodology a different set of experiments are performed sequentially. It uses initially the first-order design. The most common first-order design consists of n_f two-level factorial design points augmented with n_c central points. The n_f factorial design points are also called corner points. The factorial design may be full or fractional depending on what level of resolution discussed in Section 6.5.1 is required. The experiments are replicated only at a central point to estimate system/process error. The number of replications is chosen in such a way that it provides stable variance of predicted response. The optimum n_c may be found from RSM literature [5,6]. The levels are in coded form. The maximum and minimum levels are taken as $+1$ and -1 while the middle level called the central point is taken as 0. The coding formula is given in Section 6.4.1. An example of first-order design is depicted in Figure 8.1a. The solid circular points in the

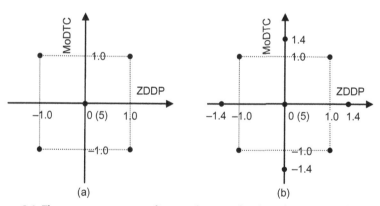

Figure 8.1 The most common first and second-order designs used in RSM. (a) Experimental points for a first-order design having 2^2 full factorial design augmented with five central points, (b) experimental points for a second-order design known as rotatable CCD having 2^2 full factorial design augmented with five central points and four axial points.

figure are the experimental points. The total number of experiments in this design is $N = n_f + n_c = 2^2 + 5 = 9$. The region of experimentation is selected in such a way so that linear model may be fitted adequately. The next set of experimentation is performed according to first-order design with a central point X_{i+1}. X_{i+1} is obtained using the following equation:

$$X_{i+1} = X_i + \lambda_{opt} S_i \qquad (8.1)$$

where

X_i, X_{i+1} = the central points at ith and $(i + 1)$th set of experiments,

λ_{opt} = optimum step length in search direction S_i,

S_i = search direction at ith set of experiments = ∇f (if objective is to maximize and is called steepest ascent direction) = $-\nabla f$ (if objective is to minimize and is called steepest descent direction).

The optimum step length value λ_{opt} in the search direction is obtained by doing several experiments at different single points $X'_{i+1} = X_i + \lambda S_i$ along search direction S_i by taking different values of λ. The value of λ at which the response is optimum is taken as λ_{opt}. It should be noted that this response is only locally optimum along search direction around central point X_i to obtain λ_{opt}. The optimum region in the entire feasible range of variables is obtained by performing a series of experiments according to first-order design in which the central point is taken as per Eqn (8.1). It also should be noted here that a new search direction is obtained from the first-order model at each set of experiments. When we reach in the optimum region the first-order model will not be applicable since there is curvature in this region. The adequacy of the model may be checked from the methods given in Section 6.6.2. At this stage a new set of experiments are performed using second-order design. The most common second-order design is CCD. This design consists of n_f two-level factorial design points augmented with n_c central points and $2k$ axial points where k is number of factors or variables. The coded levels of n_f corner points and n_c central points are same as in first-order design discussed earlier in this section. The $2k$ axial points are on the axis of variables. For each variable two axial points are there at a distance $\pm \alpha$ from the center. The value of α lies between 1 and $\sqrt{k}$. If $\alpha = 1$ then for each variable only three levels are required and is suitable for dealing qualitative factors/variables [5]. If $\alpha = 1$ and $k = 3$ then the axial point will be on the face of cube. This design is called Face Centred Cube Design. If $\alpha = \sqrt{k}$ then axial and corner points will be at same distance from the central point. Hence it is called spherical CCD. For good prediction variance of predicted response must be the same at all points that are at same distance from the center. Such a design is called rotatable design [5]. If $\alpha = (n_f)^{1/4}$ then CCD design becomes rotatable. One example of rotatable CCD is given in Figure 8.1b. The total number of experiments in this design is $N = n_f + n_c + 2k = 2^2 + 5 + 2 \times 2 = 13$. In this design experiments are performed at five levels of variables and the replications are done only at middle level 0. Hence the total number of experiments is much lower than full factorial design even if three levels are taken to express the curvature in the vicinity of optimum point. This is because replications are done at all experimental points in factorial designs. At this stage an appropriate regression model having order ≥ 2 is chosen. Then the optimum point can be obtained using the procedure given in the first paragraph of this section.

8.2.2 Example

Multivariable optimization is exemplified in this section by considering the work of Saini [7]. The work gives the systematic procedure to select the best combination of additives so that steady-state wear rate is minimum. Three additives, namely ZDDP, MoDTC, and borate ester were used to make different lubricant formulations in paraffinic base oil. Out of these, borate ester was commercially made while ZDDP and MoDTC were made in the laboratory. Though the additives synthesized in the lab may not correspond to commercial additives, they are adequate to develop methodology for studying interaction effects. The concentrations of these additives need to be optimized to have their best performance. The performance is related to their individual and interaction effects. In this work, rotatable CCD was used for synergistic study and optimization of additive concentration in the lubricant formulation. The experiments were performed in a 4-Ball machine. The details of experimentation, wear quantification, experimental design, and analysis are given in subsequent subsections.

The experiments conducted resulted in a nonlinear equation between quasi steady-state wear rate (qsw) and the concentration variables. Since the purpose was to optimize within the selected design space the optimization was done with the empirical equation. In effect no prior experimentation was done to reach the optimum zone as discussed in Section 8.2.1.

8.2.2.1 Test Set-Up and Test Method

In this work different lubricant formulations were made by using base oil as a solvent and adding additives to it according to the levels mentioned in Table 8.1. The base oil of SAE-30 grade was provided by the R&D center of Castrol India at Mumbai.

The ZDDP and MoDTC additives were made in the laboratory. The use of ZDDP has been fully discussed in Section 4.3.3. At present there is an important need to reduce the pollution due to zinc and also to reduce the catalyst poisoning effect due to phosphorous. So several soluble molybdenum compounds are being tried to reduce the concentration of ZDDP [8,9]. Use of borates with low SAP values with expected antiwear behavior are also of interest. These additives are reported to have a synergistic effect with ZDDP [10]. So in the present work combinations of these three additives have been studied.

The ZDDP was prepared by mixing a stoichiometric solution of phosphorous pentasulfide in ethanol and aqueous solution zinc acetate. A white

Table 8.1 Rotatable CCD with qsw for different experiments

Experiment no.	ZDDP concentration coded value: x_1	MoDTC concentration coded value: x_2	Borate ester concentration coded value: x_3	qsw (mm^3/h) $\times 10^{-4}$
1	1	1	1	3.28
2	-1	1	1	9.19
3	1	-1	1	3.28
4	-1	-1	1	25.99
5	1	1	-1	3.13
6	-1	1	-1	11.67
7	1	-1	-1	6.17
8	-1	-1	-1	13.34
9	0	0	0	2.09
10	0	0	0	4.03
11	0	0	0	2.66
12	0	0	0	2.78
13	0	0	0	4.25
14	0	0	0	1.99
15	1.68	0	0	2.82
16	-1.68	0	0	23.10
17	0	1.68	0	4.88
18	0	-1.68	0	7.05
19	0	0	1.68	9.08
20	0	0	-1.68	4.19

precipitate of ZDDP was obtained, which was filtered, washed with ether, and dried. The MoDTC was prepared by mixing stoichiometric solution of sodium diethyldithiocarbamate in ethanol and slightly acidic aqueous solution of sodium molybdate at around 4°C. A yellow precipitate of MoDTC was obtained, which was filtered, washed with ether, and dried. The borate ester is commercial and was obtained from R.T. Vanderbilt Company, Inc.

Each formulation was stirred for 1 h using a magnetic stirrer to make a uniform suspension of additives in the base oil. Each solution was tested on the four ball tester and all tests were performed at a load of 40 kg, 1450 rpm, and room temperature for a total test time of 90 min. The experiments were performed in a random order. The total formulations are referred to as TLF.

The wear in the 4-Ball test was monitored at 15, 30, 60, and 90 min of the test by measuring wear scar diameters in the direction of sliding and perpendicular to it on the three stationary lower balls. The wear scar diameters were measured by tilting the microscope's eye piece at an angle around 28° without disturbing the position of the balls. Mean wear scar diameter was found by using the formula in Eqn (7.2), which is the equivalent circular diameter obtained from the summation of the elliptical scar area of three balls. This value was used to obtain the wear volume from Eqn (4.11), which takes elastic recovery into account.

8.2.2.2 Methodology to Find Quasi Steady-State Wear Rate

As discussed in Section 7.4.2, the procedure for wear comparison was based on the wear rate obtained from the slope at 90 min. This is because the test durations are relatively short, unlike the reciprocating tests run for a long duration of 8 h. In the case of long duration tests the procedure adapted was given in Section 7.3.1. The wear rate at 90 min was characterized by taking the slope at 90 min on a nonlinear regressed curve between wear volume versus time. The curve was obtained from the computer program developed by Kumar et al. [11]. This wear rate was considered as qsw for comparison purposes. The qsw values are tabulated in Table 8.1 and are expressed in mm^3/h. *The computer program to obtain the qsw is annexed.*

It is important to note that as per the wear terminology discussed in Chapter 3, wear should be expressed in terms of wear volume/unit sliding distance or by nondimensional wear coefficient. The comparison on the basis of wear volume/unit time used here is valid because the sliding speed is constant.

8.2.2.3 Experimental Design

TLF experimentation was done by using a rotatable CCD. The design with coded values of additive concentrations along with qsw as a response is given in Table 8.1.

In the present CCD 2^3 factorial design is augmented by six central and six axial points. For axial points α was taken as $(n_f)^{1/4} = 8^{1/4} = 1.68$ to make the design rotatable, which was discussed in Section 8.2.1. A total of 20 experiments were conducted. The coded values of additive concentration can be converted into actual additive concentration % (w/v) by using Eqn (6.14). The additive concentration in % (w/v) at extreme and central points were taken as $z_{max} = 2\%$, $z_{min} = 0\%$, and $z_0 = 1\%$.

8.2.2.4 Statistical Analysis and Optimization

To model the qsw as a function of additive concentration a quadratic polynomial relation was assumed, in which all combinations of two factor interaction terms were taken into account. The coefficients of the polynomial were found by using Eqn (6.23), in which matrices X and Y were as shown here:

$$
X = \begin{bmatrix}
1 & 1 & 1 & 1 & 1 & 1 & 1 & 1 & 1 & 1 \\
1 & -1 & 1 & 1 & 1 & 1 & 1 & -1 & -1 & 1 \\
1 & 1 & -1 & 1 & 1 & 1 & 1 & -1 & 1 & -1 \\
1 & -1 & -1 & 1 & 1 & 1 & 1 & 1 & -1 & -1 \\
1 & 1 & 1 & -1 & 1 & 1 & 1 & 1 & -1 & -1 \\
1 & -1 & 1 & -1 & 1 & 1 & 1 & -1 & 1 & -1 \\
1 & 1 & -1 & -1 & 1 & 1 & 1 & -1 & -1 & 1 \\
1 & -1 & -1 & -1 & 1 & 1 & 1 & 1 & 1 & 1 \\
1 & 0 & 0 & 0 & 0 & 0 & 0 & 0 & 0 & 0 \\
1 & 0 & 0 & 0 & 0 & 0 & 0 & 0 & 0 & 0 \\
1 & 0 & 0 & 0 & 0 & 0 & 0 & 0 & 0 & 0 \\
1 & 0 & 0 & 0 & 0 & 0 & 0 & 0 & 0 & 0 \\
1 & 0 & 0 & 0 & 0 & 0 & 0 & 0 & 0 & 0 \\
1 & 0 & 0 & 0 & 0 & 0 & 0 & 0 & 0 & 0 \\
1 & 1.68 & 0 & 0 & 2.82 & 0 & 0 & 0 & 0 & 0 \\
1 & -1.68 & 0 & 0 & 2.82 & 0 & 0 & 0 & 0 & 0 \\
1 & 0 & 1.68 & 0 & 0 & 2.82 & 0 & 0 & 0 & 0 \\
1 & 0 & -1.68 & 0 & 0 & 2.82 & 0 & 0 & 0 & 0 \\
1 & 0 & 0 & 1.68 & 0 & 0 & 2.82 & 0 & 0 & 0 \\
1 & 0 & 0 & -1.68 & 0 & 0 & 2.82 & 0 & 0 & 0
\end{bmatrix}, \quad
Y = \begin{bmatrix}
3.28 \times 10^{-4} \\
9.19 \times 10^{-4} \\
3.28 \times 10^{-4} \\
2.60 \times 10^{-3} \\
3.13 \times 10^{-4} \\
1.17 \times 10^{-3} \\
6.17 \times 10^{-4} \\
1.33 \times 10^{-3} \\
2.09 \times 10^{-4} \\
4.03 \times 10^{-4} \\
2.66 \times 10^{-4} \\
2.78 \times 10^{-4} \\
4.25 \times 10^{-4} \\
1.99 \times 10^{-4} \\
2.82 \times 10^{-4} \\
2.31 \times 10^{-3} \\
4.88 \times 10^{-4} \\
7.05 \times 10^{-4} \\
9.08 \times 10^{-4} \\
4.19 \times 10^{-4}
\end{bmatrix}
$$

In X matrix of order 20×10 the first column is for the constant term; the second to fourth columns are for first-order terms x_1, x_2, and x_3; the fifth to seventh columns are for second-order terms x_1^2, x_2^2, and x_3^2;

the eighth to tenth columns are for interaction terms x_1x_2, x_1x_3, and x_2x_3. The Y column matrix gives qsw at a different combination of additive concentrations. The coefficient matrix b obtained from Eqn (6.23) for the above model is given below.

$$b = [2.95 \quad -5.75 \quad -1.84 \quad 1.15 \quad 3.68 \quad 1.20 \quad 1.43 \quad 1.93 \quad -1.61 \quad -1.51]^T \times 10^{-4}$$

The empirical equation obtained from the above procedure is given below.

$$qsw = \left(\begin{matrix} 2.95 - 5.75x_1 - 1.84x_2 + 1.15x_3 + 3.68x_1{}^2 + 1.20x_2{}^2 + 1.43x_3{}^2 \\ + 1.93x_1x_2 - 1.61x_1x_3 - 1.51x_2x_3 \end{matrix} \right) \times 10^{-4}$$

$$(8.2)$$

Variance analysis was carried out in which the significance of the model as a whole and the significance of the individual terms in the model were tested by following the procedure described in Section 6.6.1. The F-values for different factors are tabulated in Table 8.2 and their significance is also specified. The relevant MATLAB program is annexed.

Table 8.2 ANOVA for the second-order model fitted to the qsw

Source	Sum of squares ($\times 10^{-8}$)	df	Mean square ($\times 10^{-8}$)	F-ratio	$F_{0.05,\nu_1,\nu_2}$	Significance
Model	802.44	9	89.16	14.21	$F_{0.05,9,10} = 3.02$	Significant
x_1	450.47	1	450.47	71.81	$F_{0.05,1,10} = 4.96$	Significant
x_2	46.38	1	46.38	7.39	$F_{0.05,1,10} = 4.96$	Significant
x_3	17.94	1	17.94	2.86	$F_{0.05,1,10} = 4.96$	Not significant
x_1x_2	29.76	1	29.76	4.74	$F_{0.05,1,10} = 4.96$	Not significant
x_1x_3	20.83	1	20.83	3.32	$F_{0.05,1,10} = 4.96$	Not significant
x_2x_3	18.27	1	18.27	2.91	$F_{0.05,1,10} = 4.96$	Not significant
$x_1{}^2$	194.07	1	194.07	30.94	$F_{0.05,1,10} = 4.96$	Significant
$x_2{}^2$	20.59	1	20.59	3.28	$F_{0.05,1,10} = 4.96$	Not significant
$x_3{}^2$	29.56	1	29.56	4.71	$F_{0.05,1,10} = 4.96$	Not significant
Residual	62.73	10	6.27			
Lack of fit	58.10	5	11.62	12.55	$F_{0.05,5,5} = 5.05$	Significant
Pure error	4.63	5	0.93			
Total error	865.17	19				

The model is found to be significant in which the reference is residual. The lack of fit was found to be significant in which the reference is pure error found by replication. The significance of lack of fit implies that a more appropriate model can be found by including some higher order terms. This can be done by trial and error. Higher order terms can be included in the model, which is expected to be significant based on the practical knowledge. Then again a new model is fitted on the available data and ANOVA is performed on the new model. Lack of fit is calculated each time a new term is added until it becomes insignificant.

The coefficient of determination R^2 using Eqn (6.30) was found to be 0.93. By adding higher order terms whether it is significant or not the R^2 value will increase but it does not mean that model prediction is good. For good prediction, the adjusted coefficient of determination R^2_{adj} given in Eqn (6.31) should be estimated. The value obtained is 0.86 for the data in Table 8.2. This value is quite low compared to the value of R^2. The large difference in R^2 and R^2_{adj} implies that nonsignificant terms are involved in the model. This clearly can be seen from Table 8.2.

From this analysis it is clear that some higher order terms should be added in the model for better prediction. It was found that when $x_1x_2x_3$ and $x_1^2x_2$ are included in the model the lack of fit became insignificant. The model obtained by adding these terms is given below.

$$qsw = \begin{pmatrix} 2.95 - 5.75x_1 - 0.65x_2 + 1.15x_3 + 3.68x_1{}^2 + 1.20x_2{}^2 + 1.43x_3{}^2 \\ + 1.93x_1x_2 - 1.61x_1x_3 - 1.51x_2x_3 + 2.27x_1x_2x_3 - 2.04x_1{}^2x_2 \end{pmatrix} \times 10^{-4}$$

(8.3)

The variance analysis was carried out as done in the previous model, which is summarized in Table 8.3.

The model was found to be significant while the lack of fit was found to be insignificant. Thus the model is good. R^2 and R^2_{adj} values for the above model were found to be 0.99 and 0.98, respectively. These values were calculated from Eqns (6.30) and (6.31), respectively. The larger value of R^2_{adj} implies that the prediction is better than the previous model. The difference in R^2 and R^2_{adj} is small, hence nonsignificant terms are less in the model, which also can be seen from Table 8.3 where only x_2 was found to be insignificant. In this model the terms in decreasing order of significance are x_1, x_1^2, $x_1x_2x_3$, x_1x_2, x_3^2, x_1x_3, x_2^2,

Table 8.3 ANOVA table for the qsw model including three degree terms

Source	Sum of squares ($\times 10^{-8}$)	df	Mean square ($\times 10^{-8}$)	F-ratio	$F_{0.05,\nu_1,\nu_2}$	Significance
Model	857.52	11	77.96	81.55	$F_{0.05,11,8} = 3.32$	Significant
x_1	450.47	1	450.47	471.24	$F_{0.05,1,8} = 5.32$	Significant
x_2	2.35	1	2.35	2.46	$F_{0.05,1,8} = 5.32$	Not significant
x_3	17.94	1	17.94	18.77	$F_{0.05,1,8} = 5.32$	Significant
x_1x_2	29.76	1	29.76	31.13	$F_{0.05,1,8} = 5.32$	Significant
x_1x_3	20.83	1	20.83	21.79	$F_{0.05,1,8} = 5.32$	Significant
x_2x_3	18.27	1	18.27	19.11	$F_{0.05,1,8} = 5.32$	Significant
x_1^2	194.07	1	194.07	203.02	$F_{0.05,1,8} = 5.32$	Significant
x_2^2	20.59	1	20.59	21.54	$F_{0.05,1,8} = 5.32$	Significant
x_3^2	29.56	1	29.56	30.93	$F_{0.05,1,8} = 5.32$	Significant
$x_1x_2x_3$	41.27	1	41.27	43.17	$F_{0.05,1,8} = 5.32$	Significant
$x_1^2x_2$	13.81	1	13.81	14.45	$F_{0.05,1,8} = 5.32$	Significant
Residual	7.65	8	0.96			
Lack of fit	3.02	3	1.01	1.09	$F_{0.05,3,5} = 5.41$	Not significant
Pure error	4.63	5	0.93			
Total error	865.17	19				

x_2x_3, x_3, and $x_1^2x_2$. This clearly indicates that there is interaction between additives. The positive or negative interaction can be identified by signs of the coefficient of the term and of coded values of concentration of ZDDP, MoDTC, and borate ester. The coded values vary from -1.68 to 1.68. The lowest -1.68 coded value represents 0% (w/v) while highest $+1.68$ coded value represents 2% (w/v). The coefficient of three-factor interaction is $+2.27$ but if additive concentrations are chosen in such a way that the product of coded values is negative then it shows synergy between these additives. Similarly synergy between ZDDP and MoDTC may be observed. The synergy can also be observed with response surface, which has been shown after the following discussion.

The next step is optimization. The optimum combination of additive concentration was found by using the Solver Add-in tool of MS Excel 7 in model Eqn (8.3), in which the Newton method was selected for search direction. The optimum concentrations for ZDDP, MoDTC, and borate ester in the coded form from this model came out to be $\begin{bmatrix} 0.82 & 0.18 & 0.04 \end{bmatrix}^T$. The qsw at these levels was 6.67×10^{-5} mm^3/h.

The steps to use the MS Excel 7 solver tool to obtain these results are given below:

1. If the Solver Add-in tool is not added in MS Excel it should be added following the procedure as given below:

    a. Open MS Excel and select the buttons in the following order: Office → Excel Options → Add-ins → Go.

    b. The Add-in dialog box is now opened. Choose Solver Add-in and then select OK.

    The above procedure can be followed in the MS Excel 7 and 10 versions. For the MS Excel 3 version select Tools button and then choose Add-in. After that, follow step (b).

2. Open MS Excel and choose an initial trial solution. In the present case the central point $\begin{bmatrix} 0 & 0 & 0 \end{bmatrix}^T$ was chosen as an initial solution and filled in A1, A2, and A3 cells. Figure 8.2a shows the location of filling the initial solution. A1, A2, and A3 cells have been named ZDDP_concentration, MoDTC_concentration, and borate_ester_concentration by right-clicking and choosing Name a Range.... These cells are called Adjustable Cells. Care should be taken in the selection of an initial solution if the objective function is not unimodal. The unimodal function has only one optimum point while the multimodal function has more than one optimum point. There are chances of getting a solution as relative optima rather than global optimum point if the proper initial solution is not taken. The details may be found in optimization literature [1].

3. Select a cell as the target cell. In the present case the B1 cell has been selected as the target cell and named quasi_steady_state_wear_rate. Select the target cell and write the regressed equation in the Formula Bar to the right of fx as shown in Figure 8.2a and enter it. Entering can also be done by selecting the $\sqrt{}$ button located just before fx. The B1 cell now shows the current qsw 2.95×10^{-4} at the initial trial solution $\begin{bmatrix} 0 & 0 & 0 \end{bmatrix}^T$.

4. Click the Data and then Solver buttons. A Solver Parameters dialog box appears (Figure 8.2a). Set Target Cell as B1. Choose Min since the objective is to minimize the qsw. The By Changing Cells bar should be filled as $A$1:$A$3. This can also be done by selecting A1 to A3 cells. The value of these cells changes at each iteration. Add the constraints by clicking the Add button. The Add Constraints dialog box appears. Click in Cell Reference and select A1 cell, then click in the Constraint: box and write 1.68 after choosing $<=$ in between

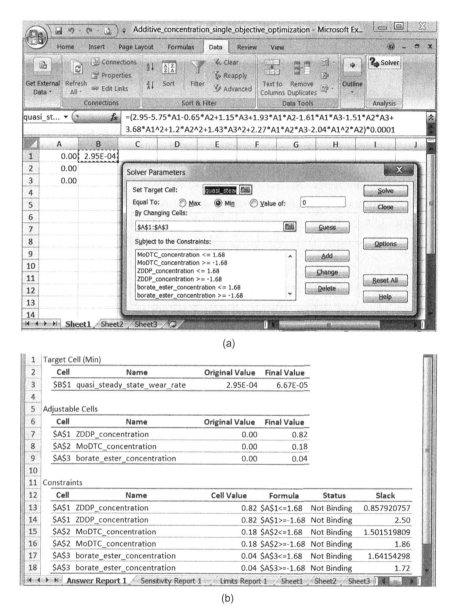

(a)

(b)

Figure 8.2 Optimization solution using MS Excel (a) initial trial solution as $[0\ \ 0\ \ 0]^T$ filling in cells A1, A2, and A3 and objective function in cell B1; (b) answer report after nine iterations showing original and final values of target and adjustable cells along with details of constraints.

these boxes. This operation will enter the constraint of ZDDP concentration $\leq = 1.68$. In this way add all six constraints as given in Figure 8.2a.

5. Click the options button. The Solver Options dialog box appears. The default values of Max time of solution, number of iterations, precision required, tolerance %, and convergence criteria are specified. These default values may be altered if required. Iteration results can be viewed after selecting the Show Iteration Results button. The default selection of estimates, derivative finding method, and search direction are Tangent, Forward, and Newton, respectively. The other options available may be selected depending on the type of optimization problem. The more versatile method, GRG nonlinear, can be selected in a new version of MS Excel 10. After selecting options click OK. The Solver Parameters dialog box appears again.

6. Click Solve. The Show Trial Solution dialog box appears. It gives the current solution at a particular iteration. Click Continue. After nine iterations the Solver Results dialog box appears. Choose Answer, Reports, and Limits. Click OK. The Answer Report 1, Sensitivity Report 1, and Limits Report 1 buttons are generated. The results can be seen by clicking Answer Report 1. Figure 8.2b shows the initial and final values of the variables ZDDP, MoDTC, and borate ester concentrations as well as the objective function qsw. The Sensitivity report gives all three elements of ∇f zero, hence we have reached the optimum point. The Limits report is useful if the constraints are there.

The qsw variation as a function of coded values of additive concentrations of ZDDP and MoDTC for fixed optimum coded value of borate ester concentration of 0.04 is given in Figure 8.3.

Note, however, that in the figure steady-state wear rate actually refers to qsw. The contours of constant qsw are also shown in the figure. In this figure if we move along the ZDDP concentration axis from -1.68 to 1.68 coded value at MoDTC concentration of -1.68 coded value (i.e., 0%, w/v) a decrease in qsw is observed initially and then increases. Moving along the MoDTC concentration axis from -1.68 to 1.68 coded value at ZDDP concentration of -1.68 coded value (i.e., 0%, w/v) a decrease in qsw is observed within the experimental design space. If we move along an optimum point between these two boundaries starting from -1.68 coded value the sharper decrease in qsw can be observed. The minimum qsw is at ZDDP and MoDTC concentrations of 0.82 and 0.18 coded values. It clearly shows that there is synergy between ZDDP

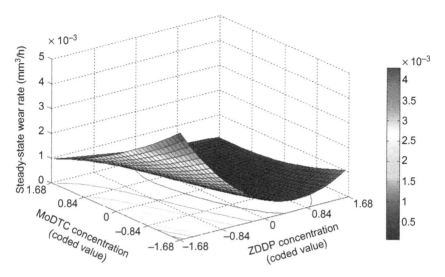

Figure 8.3 qsw variation as a function of ZDDP and MoDTC concentration in coded value.

and MoDTC. Similarly synergy between other additive interactions can be viewed.

The main conclusions drawn from this discussion are as follows:

1. On the basis of designed experiments it can be concluded that the main effect of ZDDP was found to be the most significant parameter having F ratios of 471.24 and 203.02 for linear and quadratic concentration terms in the model for qsw.

2. The optimum combination was found to be 1.49% (w/v) for ZDDP, 1.11% (w/v) for MoDTC, and 1.02% (w/v) for borate ester and the corresponding qsw was found to be 6.67×10^{-5} mm^3/h.

3. The systematic statistical approach for synergistic study and optimization may be used in developing effective formulations.

8.3 MULTIOBJECTIVE OPTIMIZATION

8.3.1 Mathematical Approach

The multiobjective optimization problem can be solved by many methods available in the literature [1,2] by converting the problem into a single objective and then solving it by following the procedure described in Section 8.2. Using these methods all responses are optimized simultaneously. Some popular methods of formulation with a single objective are:

1. Develop an overall objective function by taking a linear combination of the different objective functions involved. Let $f_1(X)$ and $f_2(X)$ are two objective functions representing minimization of qsw and maximization of load carrying capacity. The overall objective function may be written as $f(X) = w_1 f_1(X) - w_2 f_2(X)$ where w_1 and w_2 are weighting factors corresponding to objective functions 1 and 2, respectively. The sum of these factors is 1. The negative sign associated with the second objective function converts the maximization problem into a minimization problem. The overall objective function so obtained may now be minimized.

2. If the target values of different objective functions $f_i(X)$ are assigned as T_i the overall objective function may be defined as the sum of the square of the difference between the objective function value and the target value. Mathematically it can be expressed as $f(X) = \sum_{i=1}^{m} (f_i(X) - T_i)^2$ where m is the number of objective functions. The minimization of the function $f(X)$ will give each objective function values closer to the target values.

3. Assign the most important objective function as the main objective function and the other as constraints representing the acceptable range of the individual objective function. Now the formulated problem is called the constraint nonlinear optimization problem, which may be solved by using methods described in Section 8.2.1 or by using MS Excel.

4. Transform each objective function $f_i(X)$ into the desirability function $d_i(X)$ [12], the value of which lies between 0 and 1. If the objective function value meets the target then the desirability function value is 1 and if it is out of range then $d_i(X)$ is zero. The different types of transformation are given in Section 8.3.1.1. After converting all objective functions into desirability functions an overall desirability function D is obtained by taking the geometric mean of all $d_i(X)$ expressed as $D = \left(\prod_{i=1}^{m} d_i(X)\right)^{1/m}$, where m is the number of objective functions. D is now maximized. The D value gives the overall desirability of all objectives and value lies between 0 and 1. If any objective function lies outside the range (i.e., $d_i(X)$ is zero) then D will also be zero. For example, if a particular response of lubricant performance is unacceptable then that lubricant may be rejected even if other responses give a better desirability value. This is the reason why the geometric mean of d_i values is taken instead of the arithmetic mean.

8.3.1.1 Different types of transformations for converting objective function into desirability function

Two types of transformations are used to convert an objective function into a desirability function: one-sided and two-sided. One-sided is used if the target is to have a maximum or minimum possible value T of $f(X)$. In these cases the transformation is on one side of the target value. Two-sided transformation is used if the target is to have nominal value T of $f(X)$. In this case T lies between two constraints and transformation is both sides of target value T. The following equations are used for this purpose:

$$d_i(X) = \begin{cases} 0 & f(X) < L \\ \left| \dfrac{f(X)-L}{T-L} \right|^r & L \leq f(X) \leq T \quad \text{if } T \text{ is maximum possible value of } f(X) \\ 1 & f(X) > T \end{cases}$$

$$\text{(8.4)}$$

$$d_i(X) = \begin{cases} 1 & f(X) < T \\ \left| \dfrac{f(X)-U}{T-U} \right|^r & T \leq f(X) \leq U \quad \text{if } T \text{ is minimum possible value of } f(X) \\ 0 & f(X) > U \end{cases}$$

$$\text{(8.5)}$$

$$d_i(X) = \begin{cases} 0 & f(X) < L \\ \left| \dfrac{f(X)-L}{T-L} \right|^r & L \leq f(X) \leq T \\ \left| \dfrac{f(X)-U}{T-U} \right|^r & T \leq f(X) \leq U \\ 0 & f(X) > U \end{cases} \quad \text{if } T \text{ is nominal value of } f(X) \quad \text{(8.6)}$$

where

$L, U = $ lower and upper limit of objective function $f(X)$,

$T = $ target value of $f(X)$,

$r = $ an exponent.

The value of r quantifies the emphasis on the target value. If $r > 1$ is chosen it means the objective function value is closer to the target value. In this case the desirability value accelerates with the increase of $f(X)$ as

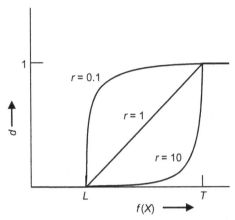

Figure 8.4 Transformation of objective function into desirability function for various values of r in the case of maximization problem.

we approach target value. If $0 < r < 1$ is chosen the above criteria is not so critical and any value within the range is acceptable. In this case the desirability value increases at a decreasing rate. If $r = 1$ linear transformation is observed. This can be viewed in Figure 8.4 for the case of the maximization problem.

8.3.2 Example

Multiobjective optimization is exemplified in this section by considering the work of Rajkamal [13]. The work gives the systematic procedure using the desirability function to select the best combination of additives so that qsw is minimum while load carrying capacity is maximum. Two additives, ZDDP and borate ester, were used to make different lubricant formulations in paraffinic base oil. Both additives were commercially made. The experiments were performed in a 4-Ball machine. The details of experimentation, experimental design, and analysis are given in subsequent subsections.

8.3.2.1 Test Set-Up and Test Method

In this work different lubricant formulations were made by using base oil as a solvent and adding additives to it according to the levels mentioned in Table 8.4. The base oil was provided by the R&D center of Castrol India at Mumbai. ZDDP and borate ester in solution form were supplied by Afton Chemicals and R.T. Vanderbilt, respectively.

Table 8.4 qsw and load carrying capacity at different experiments

Experiment no.	ZDDP concentration coded value: x_1	Borate ester concentration coded value: x_2	qsw (mm³/h) $\times 10^{-4}$	Load carrying capacity (N)
1	1	1	2.04	2000
2	-1	1	2.30	1700
3	1	-1	1.34	1900
4	-1	-1	6.72	1400
5	0	0	2.61	1400
6	0	0	1.33	1100
7	0	0	1.84	1500
8	0	0	1.70	1100
9	0	0	1.91	1100
10	1.41	0	1.54	1900
11	-1.41	0	11.80	1600
12	0	1.41	1.51	1300
13	0	-1.41	0.81	1300

TLF experimentation for the antiwear study was done as per the set-up and methods given in Section 8.2.2.1 for the multivariable optimization example. For the load carrying capacity study the peak load was determined by observing the sudden increase in the coefficient of friction in a friction test as per ASTM 5183, in which step loading was done. As per the standard, initially a 1 h test is done in a 4-Ball machine at 40 kgf load and 75°C temperature. The wear scar diameter after this test must be less than 0.67 ±0.03 mm as per the standard. After this test oil used was thrown away from the cup. Then a friction test on the worn scar with fresh formulated oil was performed, where step loading with the step of 10 kgf load was done at intervals of 10 min until the friction value shoots, indicating the beginning of seizure.

8.3.2.2 Experimental Design

TLF experimentation was done by using rotatable CCD. In the present CCD 2^2 factorial design is augmented by five central and four axial points. For axial points α was taken as $(n_f)^{1/4} = 4^{1/4} = 1.41$ to make the design rotatable, which was discussed in Section 8.2.1. Thirteen experiments were conducted for each of two responses. The design with coded values of additive concentrations along with qsw and load carrying capacity as responses is given in Table 8.4. The qsw values were obtained by following the methodology given in Section 8.2.2.2. The coded values of additive concentration can be converted into actual additive concentration % (w/v) by using Eqn (6.14). The additive concentration in % (w/v) at extreme and central points for ZDDP was taken as $z_{max} = 1.5\%$, $z_{min} = 0\%$, and $z_0 = 0.75\%$. For borate ester the value of z_{max}, z_{min}, and z_0 were 3.0%, 0%, and 1.5%, respectively.

8.3.2.3 Statistical Analysis

The equations obtained for qsw (mm^3/h) and load carrying capacity W (N) as a function of coded values of additive concentration are given below. These are obtained by the procedure given in Section 8.2.2.4.

$$qsw = \begin{pmatrix} 1.88 - 3.63x_1 + 0.25x_2 + 2.19x_1^2 - 0.56x_2^2 + 1.28x_1x_2 \\ -1.18x_1^2x_2 + 2.22x_1x_2^2 \end{pmatrix} \times 10^{-4}$$

(8.7)

$$W = 1240 + 153.03x_1 + 50x_2 + 311.25x_1^2 + 86.25x_2^2 - 50x_1x_2 \quad (8.8)$$

Variance analysis was carried out in which the significance of the model as a whole and significance of the individual terms in the model were tested by following the procedure described in Section 6.6.1 and exemplified in Section 8.2.2.4. Both the models described in Eqns (8.7) and (8.8) were found to be significant while lack of fit was found to be insignificant, thus the models are good. R^2 and R_{adj}^2 values for the qsw model were found to be 0.98 and 0.95, respectively, while for the load carrying capacity model the values were 0.76 and 0.58, respectively. In the qsw model the order of significant terms is $x_1 > x_1^2 > x_1x_2 > x_1x_2^2$. This shows that ZDDP concentration and its interaction with borate ester have a strong influence on qsw. In the load carrying capacity model only x_1^2 was found to be significant, which shows that ZDDP concentration nonlinearly increases the load carrying capacity. This conclusion has to be rechecked by doing careful controlled experiments because there is large variation between R^2 and R_{adj}^2 values of the load carrying capacity model.

The multiobjective optimization was carried out using the desirability function as described in Section 8.3.1. Desirability d_1 for the qsw model was expressed as a function of X by using Eqn (8.5) since the objective is to minimize qsw. In this target value T was taken as the minimum value obtained from considering qsw as a single objective problem. In this case the minimum value was 3.23×10^{-5} mm^3/h at the coded values of ZDDP and borate ester concentrations were 0.77 and -0.23, respectively.

These values can be obtained by using MS Excel as exemplified in Section 8.2.2.4. The upper limit U was taken as the maximum experimental value $1.18 \times 10^{-3} \, mm^3/h$ in the design space. The exponent r was taken as 1 considering linear transformation.

Desirability value d_2 for the load carrying capacity model was expressed as a function of X using Eqn (8.4) since the objective is to maximize the load carrying capacity. In this target value T was taken as the maximum value obtained in the design space considering load carrying capacity as a single objective problem. In this case the maximum value was 2275 N at the coded values of ZDDP and borate ester concentrations as 1.41 and -1.41, respectively. These values can be obtained by observing the response surface and using MS Excel as exemplified in Section 8.2.2.4. The lower limit L was taken as the minimum experimental value 1100 N in the design space. The exponent r was taken as 1 considering linear transformation.

The overall desirability D is expressed as the geometric mean of d_1 and d_2, which is a function of $X = \begin{bmatrix} x_1 & x_2 \end{bmatrix}^T$. Now D was taken as an objective function to be maximized. The range of additives concentration in design space from -1.414 to 1.414 was taken as constraints. The solution was obtained by using MS Excel's Solver Add-in tool. The Newton method was selected for the search direction. The optimum overall desirability D was found to be 0.88 at $X = \begin{bmatrix} 1.414 & 0.022 \end{bmatrix}^T$. X vector represents the coded values of the concentrations for ZDDP, and borate ester, respectively. At this optimum point the individual desirability for qsw and load carrying capacity were found to be 0.930 and 0.833, respectively, while the values of qsw and load carrying capacity were obtained as $1.13 \times 10^{-4} \, mm^3/h$ and 2078.28 N, respectively. The steps to use MS Excel's solver tool to obtain these results are given below:

1. The initial solution was chosen as the central point $\begin{bmatrix} 0 & 0 \end{bmatrix}^T$ and filled in B3 and B4 cells. Figure 8.5a shows the location of filling the initial solution.

2. Select the B7 cell and write Eqn (8.7) in the Formula bar for the quasi steady-state wear rate. Select the B8 cell and write Eqn (8.8) in the Formula Bar for load carrying capacity.

3. Select the B10 cell and write $= ABS((B7-0.00118)/(0.0000323-0.00118))$ in the Formula bar for the transformation function of the quasi steady-state wear rate. Select the B11 cell and write $= ABS((B8-1100)/(2275-1100))$ in the Formula bar for the transformation function of load carrying capacity.

4. Select the B13 cell and write $= IF(B7 > 0.00118,0,IF(B7 > = 0.0000323,B10,1))$ for the desirability value d_1 of the quasi steady-state wear rate at the current iteration. Select the B14 cell and write $= IF(B8 > 2275,1,IF(B8 > = 1100,B11,0))$ for the desirability value d_2 of load carrying capacity at the current iteration.

5. Select the E3 cell as the target cell and write $= (B13*B14)^{(1/2)}$ in the Formula bar for overall desirability D. The E3 cell now shows the current overall desirability 0.321 at the initial trial solution $\begin{bmatrix} 0 & 0 \end{bmatrix}^T$.

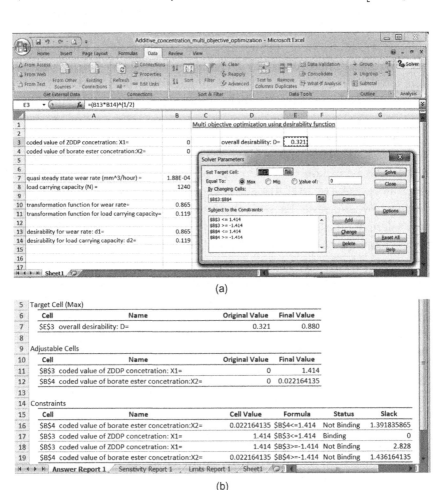

(a)

(b)

Figure 8.5 Multiobjective optimization solution using MS Excel. (a) Initial trial solution as $\begin{bmatrix} 0 & 0 \end{bmatrix}^T$ filling in cells B3, and B4 and objective function in cell E3. (b) Answer report after four iterations showing original and final values of target and adjustable cells along with details of constraints.

6. Click Data and then Solver. A Solver Parameters dialog box appears (Figure 8.5a). Set Target Cell as E3. Choose Max since the objective is to maximize the overall desirability. Fill the By Changing Cells bar with $B$3:$B$4. The value of these cells changes at each iteration. Add the constraints $B$3 < = 1.414, $B$3 > = −1.414, $B$4 < = 1.414, and $B$3 > = −1.414 as shown in Figure 8.5a.

7. Follow steps 5 and 6 given in Section 8.2.2.4 for running the program. After four iterations result appears. Figure 8.5b shows the initial and final values of the variables ZDDP and borate ester concentrations, as well as of the objective function: overall desirability. The details of constraints are also shown in Figure 8.5b.

Note that many DOE software programs are available to solve the optimization problem. Design Expert and DOE + + are widely used in these areas. Further it should be mentioned here that lubricant formulation on the basis of optimum solution may not be robust. Robustness can be checked by doing post-optimality analysis, in which sensitivity of change in responses is observed on changing variables; for example, concentration of additives about the optimum point [6]. The optimum region in which change in responses is small may be selected for application. In order to make lubricant formulations robust, Robust Parameter Design developed by Taguchi [14] will be very useful. In this detailed system-specific series of experimentations are required. The system-specific wear mapping mentioned in this book may be very useful in this regard. Robust lubricant formulation can also be achieved by another Response Surface Approach [6]. The optimization is a vast area for which specialized knowledge is required. Only the main concepts involved are mentioned here. The relevance of these concepts with practice is discussed in the next section.

8.4 RELEVANCE TO PRACTICE

This brief section considers the problem of DOE and optimization from the point of view of lubricant formulators. As discussed at several places in the book, wear evaluation should be based on the proposed methods. This is an important step and will enable effective comparisons between different formulations. Two aspects of relevance discussed here are physical significance of the equations and how the knowledge base available with industry can be used in developing focused methodologies.

8.4.1 Physical Significance

The polynomial equations developed on the basis of DOE will give a good idea of the various interactions. Equation (8.3) may now be examined again.

$$\text{qsw} = \left(\begin{array}{l} 2.95 - 5.75x_1 - 0.65x_2 + 1.15x_3 + 3.68x_1^2 + 1.20x_2^2 + 1.43x_3^2 \\ + 1.93x_1x_2 - 1.61x_1x_3 - 1.51x_2x_3 + 2.27x_1x_2x_3 - 2.04x_1^2x_2 \end{array} \right) \times 10^{-4}$$

On the basis of ANOVA (Table 8.3) the relative significance of each term can be obtained. For example, based on F ratios the individual effects can be rated as ZDDP > MoDTC > borate ester. Also it can be seen that the linear and quadratic components of ZDDP concentration have a strong influence on qsw compared to the other additives. This is one way of obtaining the relative importance of additives and their interactions. But when polynomial terms with three or more level interactions become significant there can be problems in interpreting the physical significance. Some other methods to test significance are also available as discussed in Section 6.2.4 but are not considered here.

8.4.2 Adapting to the Industry

The best way to follow additive interactions and main effects can be by small design (say 2^2) at a selected high temperature with only two additives and then move to the three-component systems where required. The work reported here is mainly to develop methodologies and the additive combinations are not claimed to be actual designs to be used. Any studies of industrial relevance should include temperature as a variable in the range of interest to the industry. Also use of borates may be ruled out in many systems. With regard to broad selection criteria the industry with background knowledge is unlikely to test incompatible combinations. The main interest would be the relative performance of synergistic additive combinations that can be well analyzed by the procedures in this chapter. Many present day problems like partial/complete replacement of ZDDP in the engine oil can be effectively studied by the methods explained here.

The optimization procedures normally needed in lubricant industry is to stay within a band of expected performance and it may be easy enough to design on the basis of the most severe requirements. However there will be cost constraints besides ecological considerations. These problems can be solved by expressing the constraints in terms of equalities and/or inequalities.

The basic issue of correlation between the laboratory test and the real system remains a problem and was briefly considered in Section 7.4.3. In some rare cases the R&D groups may have a nonstandard test that correlates with the real system in a specific range. In such cases it will be of immediate benefit to develop a database regarding laboratory performance.

The DOE and optimization are used in the formulation of medicines and food products [15,16] that involve similar formulation problems with constraints. With regard to lubricant formulation the existing literature does not report any major effort in this direction. It is possible that some laboratories may be using these techniques in-house. We do hope that the gap is bridged by this book.

It may be pointed out that the book dealt with the difficult problem of tribology of chemical additives. Many other lubricant formulation problems like detergency and oxidation stability can also be dealt effectively with the use of DOE and optimization.

NOMENCLATURE

b	coefficient matrix of order $k \times 1$
d	desirability value
d_1, d_2	desirability value for qsw and load carrying capacity, respectively
$d_i(X)$	desirability function for ith objective function
D	overall desirability
$f(X)$	objective function
$f_1(X), f_2(X)$	objective functions for responses 1 and 2
$f_i(X)$	objective function for ith response
H	Hessian matrix of order $k \times k$
H_j	principal determinants of Hessian matrix, $1 \leq j \leq k$
H_{ij}	$= \frac{\partial^2 f}{\partial x_i \partial x_j}$, element of Hessian matrix in ith row and jth column where $1 \leq i \leq k$ and $1 \leq j \leq k$
k	number of variables
L	lower limit of objective function
m	number of objective functions
n_f	number of factorial design points
n_c	number of central points
N	total number of experiments
qsw	quasi steady-state wear rate (mm^3/h)
r	an exponent
R^2	coefficient of determination, R-squared
R^2_{adj}	adjusted R-squared

S_i	search direction at ith set of experiments, ∇f is called steepest ascent direction, and $-\nabla f$ is called steepest descent direction
T	target value of objective function
T_i	target value of ith objective function
U	upper limit of objective function
w_1, w_2	weighting factors corresponding to objective functions 1 and 2
W	load carrying capacity (N)
x_1, x_2, x_3	coded values of ZDDP, MoDTC, and borate ester concentrations in a multivariable optimization study
x_1, x_2	coded values of ZDDP, and borate ester concentrations in a multiobjective optimization study
x_i	variable, where $1 \leq i \leq k$
X	a vector representing k variables
X	regressor variable matrix of order $N \times p$, where p is number of regressors including constant
X_i, X_{i+1}	the central points at ith and $(i + 1)$th set of experiments
X'_{i+1}	the point along search direction at step length λ from point X_i
X^*	stationary point
Y	column matrix of N observed responses having order $N \times 1$
z_{max}, z_{min}	maximum and minimum values of additive concentrations in % (w/v)
z_0	additive concentration in % (w/v) at central point

Greek Letters

α	coded distance of axial points from central point
λ	step length in search direction S_i
λ_{opt}	optimum step length in search direction S_i
∇f	gradient f, representing first partial derivatives of f with respect to k variables

REFERENCES

[1] Rao SS. Engineering optimization—theory and practice. 3rd ed. New Delhi: New Age International Publishers; 2010 [chapter 1, 2].

[2] Deb K. Optimization for engineering design—algorithms and examples. 2nd ed. New Delhi: PHI; 2012 [chapter 3, 4].

[3] Rao SS. Engineering optimization—theory and practice. 3rd ed. New Delhi: New Age International Publishers; 2010 [chapter 6, 7].

[4] Box GEP, Wilson KB. On the experimental attainment of optimum conditions. J R Stat Soc B 1951;13(1):1−45.

[5] Wu CFJ, Hamada M. Experiments—planning, analysis, and parameter design optimization. New York, NY: John Wiley and Sons; 2002 [chapter 9].

[6] Montgomery DC. Design and analysis of experiments. 5th ed. New York, NY: John Wiley and Sons; 2003 [chapter 11].

[7] Saini H. A synergistic study and optimization of additive concentrations in engine oil using DoE [M. Tech. thesis]. Varanasi: Department of Mechanical Engineering, IIT (BHU); 2010.

[8] Unnikrishnan R, Jain MC, Harinarayan AK, Mehta AK. Additive−additive interaction: an XPS study of the effect of ZDDP on the AW/EP characteristics of molybdenum based additives. Wear 2002;252:240−9.

 [9] Feng X, Jianqiang H, Junbing Y, Fazheng Z. Antiwear synergism between organomolybdenum compounds and a ZnDDP. Lub Sci 2006;18(1):1−5.
[10] http://www.vanderbiltchemicals.com/ee_content/Documents/Technical/TDS_VANLUBE_289_Web.pdf.
[11] Kumar R, Prakash B, Sethuramiah A. A systematic methodology to characterise the running-in and steady-state wear processes. Wear 2002;252:445−53.
[12] Derringer G, Suich R. Simultaneous optimization of several response variables. J Qual Technol 1980;12(4):214−19.
[13] Rabha R. Optimization of additives concentration in eco-friendly lubricant [M. Tech. thesis]. Varanasi: Department of Mechanical Engineering, IIT(BHU); 2014.
[14] Taguchi G. System of experimental design, vols 1 and 2. White Plains, NY: UNIPUB/Kraus International Publications; 1987.
[15] Mills JE. Design of experiments in pharmaceutical process research and development. In: Abdel-Magid AF, Ragan JA, editors. Chemical process research, vol. 870. Washington, DC: American Chemical Society; 2003 [chapter 6].
[16] Hu R. Food product design: a computer aided statistical approach. Boca Raton, FL: CRC Press; 1999.

ANNEXURE

The Annexure gives three MATLAB programs in Annexure-1, 2 and 3. The details of use of program are given in the following pages

ANNEXURE-1: COMPUTER PROGRAM FOR STEADY SATE WEAR RATE

This computer program is useful for long duration test. The algorithm is based on discussion in section 7.3.1. To use the program the main file 'sw. m', wear characterization function file 'wch.m', and input file 'exp_ data_sw.xls' given as follows must be copied in 'work' directory of MATLAB. MATLAB files of '.m' extension may be created by opening 'Editor' in MATLAB while Excel file exp_data_sw.xls may be created in MS-Excel. The out put is displayed on screen in a result_sw.txt file. It gives running-in wear rate, steady state wear rate, and running-in period. The output is also displayed in Figure 1 which gives experimental as well as theoretical curves between wear volume and time. To understand the algorithm the comments in the program are also given by using % sign.

Main file: sw.m

```
close all
clear all
clc;
format long;

[E] = xlsread('exp_data_sw.xls'); % reading wear volume data

n = length(E); % n is the number of experimental points including at
t = 0

x = zeros(n,2); % creating x matrix (order nx2) of zeros
y = zeros(n,1);
b = 1;

y(:,1) = E(:,2); % entering elements of y matrix by taking values of
second column of E matrix
x(:,2) = E(:,1);

[R,z,A,a,Vs,v] = wch(b,x,y,n); % returning R,z,A,a,Vs,v values by
calling wear characterizing function 'wch'

x = z;
```

213

```
% initial iteration to get b1 and b2
while (R<1)
    b2 = b;
    b = b/2;
    [R,z,A,a,Vs,v] = wch(b,x,y,n);
    x = z;
  end

b1 = b;
  if(b1 == 1)
    while(R>1)
      b1 = b;
      b = b*2;
      [R,z,A,a,Vs,v] = wch(b,x,y,n);
      x = z;
    end

b2 = b;
b = (b1 + b2)/2;

  else
    b = (b1 + b2)/2;
  end

% calculation of b using bisection method
  while(1)
    [R,z,A,a,Vs,v] = wch(b,x,y,n);
    x = z;
    q = 1-R;

    if ( q<0.0001 && q> -0.0001)
      break;

    elseif(q>0)
      b2 = b;
      b = (b1 + b2)/2;

    elseif(q<0)
      b1 = b;
      b = (b1 + b2)/2;
    end
  end
```

% calculation of running-in wear rate and running-in period

```
V0 = a*b + Vs;
p = 95;
D1 = Vs/(V0-Vs);
D2 = (100/p)-1;
tr = -log(D1*D2) /b;
```

% printing running-in wear rate, steady state wear rate, and running-in period in 'result_sw.txt' file

```
F = fopen('result_sw.txt','w');
fprintf(F,'\n running-in wear rate, V0(mm^3/hour) = %e \r\n steady
state wear rate, Vs(mm^3/hour) = %e \r\n running in period, Tr(hours) =
%f \r\n',V0,Vs,tr);
fclose(F);
```

% plotting the graph in 'Figure 1' file

```
figure(1)
grid on
hold on
xlabel ('Time (h)')
ylabel('wear (mm^3)')
plot(x(:,2),y,'b -o') %Blue- experimental
plot(x(:,2),v,'k') %Black- theoretical
hold off

winopen('result_sw.txt');
```

Wear Characterizing Function File: wch.m

```
function [R,z,A,a,Vs,v] = wch(b,x,y,n)
format long;

v = zeros(n,1);
z = zeros(n,2);

for i = 1:n
   z(i,1) = 1- exp(-b*x(i,2));
end

z(:,2) = x(:,2);
A = z\y;
a = A(1,1);
Vs = A(2,1);
```

```
for i = 1:n
  v(i,1) = a*z(i,1) + Vs*z(i,2);
end
Va = sum(v)/n;
rv = 0;
ry = 0;
for i = 1:n
  rv = rv + (v(i)-Va)^2;
  ry = ry + (y(i)-Va)^2;
end
R = sqrt(rv/ry);
```

Input file: exp_data_sw.xls

Time (t) in hours	Wear volume (V) in mm^3
0	0
0.1667	1.66E-04
0.3333	1.73E-04
0.6667	2.14E-04
1	2.25E-04
1.5	2.36E-04
2	2.64E-04
3	2.95E-04
4	3.11E-04
6	4.10E-04
8	5.18E-04

Output file: result_sw.txt

```
running-in wear rate, V0(mm^3/hour) = 1.676441e-003
steady state wear rate, Vs(mm^3/hour) = 3.986185e-005
running in period, Tr(hours) = 0.729795
```

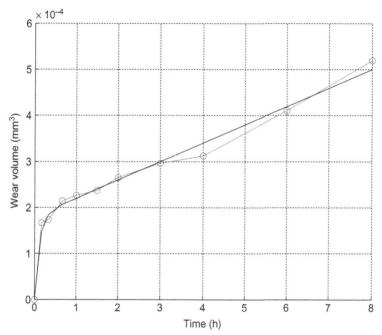

Figure 1 Output plot.

ANNEXURE-2: COMPUTER PROGRAM FOR QUASI STEADY SATE WEAR RATE

This computer program is useful for short duration test e.g. in 4-Ball machine. To use the program the main file 'qsw.m' given below, wear characterization function file 'wch.m' given in Annexure-1, and input file 'exp_data_qsw.xls' given as follows must be copied in 'work' directory of MATLAB. The method of creating files is given in Annexure-1. The output is displayed on screen in a result_qsw.txt file. It gives quasi steady state wear rate. To understand the algorithm the comments in the program are also given by using % sign.

Main file:`qsw.m`

```
close all
clear all
clc;
format long;

[E] = xlsread('exp_data_qsw.xls'); % reading wear volume data
```

```
n = length(E); % n is the number of experimental points including at
t = 0

x = zeros(n,2); % creating x matrix (order nx2) of zeros
y = zeros(n,1);

b = 1;

y(:,1) = E(:,2); % entering elements of y matrix by taking values of
second column of E matrix
x(:,2) = E(:,1);

[R,z,A,a,Vs,v] = wch(b,x,y,n); % returning R,z,A,a,Vs,v values by
calling wear characterizing function 'wch'

x = z;

% initial iteration to get b1 and b2
while (R<1)
    b2 = b;
    b = b/2;
    [R,z,A,a,Vs,v] = wch(b,x,y,n);
    x = z;
end

b1 = b;

if(b1 = = 1)
    while(R>1)
        b1 = b;
        b = b*2;
        [R,z,A,a,Vs,v] = wch(b,x,y,n);
        x = z;
    end

    b2 = b;
    b = (b1 + b2)/2;
else
    b = (b1 + b2)/2;
end

% calculation of b using bisection method
    while(1)
        [R,z,A,a,Vs,v] = wch(b,x,y,n);
        x = z;
        q = 1-R;
```

```
    if ( q<0.0001 && q> -0.0001)
       break;

    elseif(q>0)
       b2 = b;
       b = (b1 + b2)/2;

    elseif(q<0)
       b1 = b;
       b = (b1 + b2)/2;
    end
  end
```

% calculation of quasi steady state wear rate
V90 = a*b*exp(-b*1.5) + Vs; % gives wear rate at 90 minutes of test in 4-Ball machine. The equation can be derived by differentiating Eqn. (7.7) and then finding the wear rate at t = 1.5 hours.

% printing quasi steady state wear rate in 'result_qsw.txt' file

```
F = fopen('result_qsw.txt','w');
   fprintf(F,'\n quasi steady state wear rate, qsw(mm^3/hour))= %e',
V90);
   fclose(F);
winopen('result_qsw.txt');
```

Input file: exp_data_qsw.xls

Time (t) in hours	Wear volume (V) in mm^3
0	0
0.25	2.42E-04
0.5	3.09E-04
1	3.77E-04
1.5	4.09E-04

Output file: result_qsw.txt

quasi steady state wear rate, qsw(mm^3/hour)) = 8.049935e-005

ANNEXURE-3: COMPUTER PROGRAM
FOR ANOVA TABLE 8.2

This computer program generates the data for ANOVA given in Table 8.2. The program also gives the coefficients of the model alongwith R^2 and R^2_{adj} values. The program can be run by copying anova.m file in 'work' directory of MATLAB. The output is displayed on screen in a result_anova.txt file.

Main file: anova.m

```
clear
clc
X = [1 1 1 1 1 1 1 1 1 1
1 -1 1 1 1 1 1 -1 -1 1 1
1 1 -1 1 1 1 1 -1 1 -1
1 -1 -1 1 1 1 1 1 -1 -1
1 1 1 -1 1 1 1 1 -1 -1
1 -1 1 -1 1 1 1 -1 1 -1
1 1 -1 -1 1 1 1 -1 -1 1
1 -1 -1 -1 1 1 1 1 1 1
1 0 0 0 0 0 0 0 0 0
1 0 0 0 0 0 0 0 0 0
1 0 0 0 0 0 0 0 0 0
1 0 0 0 0 0 0 0 0 0
1 0 0 0 0 0 0 0 0 0
1 0 0 0 0 0 0 0 0 0
1 1.68 0 0 2.8224 0 0 0 0 0
1 -1.68 0 0 2.8224 0 0 0 0 0
1 0 1.68 0 0 2.8224 0 0 0 0
1 0 -1.68 0 0 2.8224 0 0 0 0
1 0 0 1.68 0 0 2.8224 0 0 0
1 0 0 -1.68 0 0 2.8224 0 0 0];

Y = [3.28;
    9.19;
    3.28;
    25.99;
    3.13;
    11.67;
    6.17;
    13.34;
    2.09;
    4.03;
    2.66;
```

```
  2.78;
  4.25;
  1.99;
  2.82;
 23.10;
  4.88;
  7.05;
  9.08;
  4.19]*1e-4;
A = inv(X'*X)*X'*Y;
c = inv(X'*X);

S = 0;
for i = 1:1:20

   S = S + Y(i);
end

SS_reg = A'*X'*Y - S^2/20;

SS_X1 = A(2)^2/c(2,2);
SS_X2 = A(3)^2/c(3,3);
SS_X3 = A(4)^2/c(4,4);

SS_X1X2 = A(8)^2/c(8,8);
SS_X1X3 = A(9)^2/c(9,9);
SS_X2X3 = A(10)^2/c(10,10);

SS_X11 = A(5)^2/c(5,5);
SS_X22 = A(6)^2/c(6,6);
SS_X33 = A(7)^2/c(7,7);

SS_res = Y'*Y - A'*X'*Y;
Sum_central_points = 0;
for i = 9:1:14
   Sum_central_points = Sum_central_points + Y(i);
end

y0_average = Sum_central_points/6;

SS_pe = 0;
for i = 9:1:14

   SS_pe = SS_pe + (Y(i) - y0_average)^2;
end

SS_Lof = SS_res - SS_pe;
SS_Pure_Error = SS_pe;
```

```
MS_Lof = (SS_res-SS_pe)/5;
MS_pe = SS_pe/5;

SST = Y'*Y-S^2/20;
R_square = 1-SS_res/SST;
R_square_adj = 1-(20-1)*(1-R_square)/(20-10);
MS_res = SS_res/10;
MS_reg = SS_reg/9;

F_model = MS_reg/MS_res;
F_X1 = SS_X1/MS_res;
F_X2 = SS_X2/MS_res;
F_X3 = SS_X3/MS_res;
F_X1X2 = SS_X1X2/MS_res;
F_X1X3 = SS_X1X3/MS_res;
F_X2X3 = SS_X2X3/MS_res;
F_X11 = SS_X11/MS_res;
F_X22 = SS_X22/MS_res;
F_X33 = SS_X33/MS_res;
F_Lof = MS_Lof/MS_pe;
```

% printing sum of squares, mean square, and F-ratio in 'result_anova. txt' file

```
    F = fopen('result_anova.txt','w');
    fprintf(F,'\n Coefficents\r\n');
    fprintf(F,'\n %e\r\n',A);

    fprintf(F,'\n SS_Model = %e, MS_model = %e
F_model = %f\r\n',SS_reg,MS_reg,F_model);

    fprintf(F,'\n SS_X1 = %e, MS_X1 = %e
F_X1 = %f\r\n',SS_X1,SS_X1,F_X1);

    fprintf(F,'\n SS_X2 = %e, MS_X2 = %e
F_X2 = %f\r\n',SS_X2,SS_X2,F_X2);
    fprintf(F,'\n SS_X3 = %e, MS_X3 = %e
F_X3 = %f\r\n',SS_X3,SS_X3,F_X3);
    fprintf(F,'\n SS_X1X2 = %e, MS_X1X2 = %e
F_X1X2 = %f\r\n',SS_X1X2,SS_X1X2,F_X1X2);

    fprintf(F,'\n SS_X1X3 = %e, MS_X1X3 = %e
F_X1X3 = %f\r\n',SS_X1X3,SS_X1X3,F_X1X3);
    fprintf(F,'\n SS_X2X3 = %e, MS_X2X3 = %e
F_X2X3 = %f\r\n',SS_X2X3,SS_X2X3,F_X2X3);
    fprintf(F,'\n SS_X11 = %e, MS_X11 = %e
F_X11 = %f\r\n',SS_X11,SS_X11,F_X11);
```

```
fprintf(F,'\n SS_X22 = %e, MS_X22 = %e
F_X22 = %f\r\n',SS_X22,SS_X22,F_X22);
fprintf(F,'\n SS_X33 = %e, MS_X33 = %e
F_X33 = %f\r\n',SS_X33,SS_X33,F_X33);
fprintf(F,'\n SS_residual = %e,
MS_residual = %e\r\n',SS_res,MS_res);
fprintf(F,'\n SS lack of fit = %e, MS lack of fit = %e,
F_Lof = %f\r\n',SS_Lof,MS_Lof,F_Lof);

fprintf(F,'\n SS pure error = %e, MS pure
error = %e\r\n',SS_pe,MS_pe);
fprintf(F,'\n SS total error = %e\r\n',SST);

fprintf(F,'\n R^2 = %f, R^2_adj = %f\r\n',R_square,R_square_adj);

fclose(F);
winopen('result_anova.txt');
```

Output file: result_anova.txt

```
Coefficents
2.945395e-004
-5.745808e-004
-1.843603e-004
1.146605e-004
3.675569e-004
1.197182e-004
1.434568e-004
1.928750e-004
-1.613750e-004
-1.511250e-004

SS_Model = 8.024370e-006, MS_model = 8.915967e-007
F_model = 14.213548SS_X1 = 4.504736e-006, MS_X1 = 4.504736e-006
F_X1 = 71.813060
SS_X2 = 4.637695e-007, MS_X2 = 4.637695e-007 F_X2 = 7.393264
SS_X3 = 1.793887e-007, MS_X3 = 1.793887e-007 F_X3 = 2.859757
SS_X1X2 = 2.976061e-007, MS_X1X2 = 2.976061e-007 F_X1X2 = 4.744341
SS_X1X3 = 2.083351e-007, MS_X1X3 = 2.083351e-007 F_X1X3 = 3.321212
SS_X2X3 = 1.827101e-007, MS_X2X3 = 1.827101e-007 F_X2X3 = 2.912706
SS_X11 = 1.940744e-006, MS_X11 = 1.940744e-006 F_X11 = 30.938714
SS_X22 = 2.058919e-007, MS_X22 = 2.058919e-007 F_X22 = 3.282262
SS_X33 = 2.956389e-007, MS_X33 = 2.956389e-007 F_X33 = 4.712980
SS_residual = 6.272865e-007, MS_residual = 6.272865e-008
```

SS lack of fit $= 5.809972e\text{-}007$, MS lack of fit $= 1.161994e\text{-}007$, F_Lof $= 12.551427$

SS pure error $= 4.628933e\text{-}008$, MS pure error $= 9.257867e\text{-}009$

SS total error $= 8.651657e\text{-}006$

R^2 $= 0.927495$, R^2_adj $= 0.862241$

INDEX

Printed in the United States
By Bookmasters